DECISIONS FOR SURVIVAL

Farm Management Strategies in the Middle Hills of Nepal

Jagannath Adhikari

Adroit Publishers
Delhi - 110 053

2000

DECISIONS FOR SURVIVAL

ISBN 81-87392-08-8

Price : Rs. 300/-

Published by
ADROIT PUBLISHERS
C-8/2, Yamuna Vihar,
Delhi-110053

Laser Typesetting at :
Nidhi Laser Point, Delhi.

Printed at :
New A.S. Offset Press, New Delhi.

PREFACE

This study describes and analyzes how farm families in the middle hills region of Nepal make decisions in the production of crops and in the management of private as well as common resources. It is particularly concerned with the effects of physical and socio-cultural environments and external influences (like government's interventions) on the economic behavior of farm families, including their management of local resources - land, forest, livestock and water.

This study, which was conducted in 1987 and 1993, revealed that farmers were following a diversified and integrated farming system as their tradition which, in fact, enabled them to utilize the opportunities provided by diverse nature of physical environment and to secure a part of their livelihoods from their own farm by efficiently utilizing family labor and other resources. This system also helped them to reduce the risks and uncertainties arising from natural causes like hailstorm, rain, wind, pest, insect, disease and rodents. Since the farming system followed by farm households integrated resources held in private (land and livestock) as well as in common (forest and water resources), decision-making at the village level also affected the decision-making at the family level, which was mainly concerned with the utilization of private resources. Despite the retention of fundamental elements of the diversified and integrated farming system, there were also differences in the types of crops and varieties grown across different farm-size groups. This was linked to their different goals and objectives in farming. This difference in the goals and objectives also led to conflicting situations at the village level for the management of common resources.

Farming was also intimately integrated with the family and village life of the farmers. Therefore social organization of the family and village society and cultural and religious affiliations of various groups of farmers also affected their decision-making. But the beliefs, perceptions and attitudes of farmers did not impede the adoption of modern technology. In fact it was revealed from various evidence that innovations that were suitable to the village environment and having beneficial effects on the whole farming system were successfully adopted. Even though, farmers were bounded by different kinds of obligations to various organizations and institutions, they in fact, helped in preserving the resources, mainly the ones held in common. Moreover, institutions impeding the adoption of new technology or the exploitation of new opportunities were also slowly abandoned.

Recommendations for policy making include assorted specific proposals relating to strategies for the development of technologies, extension methods and improvement of living standard.

Jagannath Adhikari
Pokhara, Nepal
July 25, 1999

ACKNOWLEDGEMENTS

This book is based on field studies conducted in 1987 and 1993 at the same research site. Accordingly, it is believed that the book can shed light on temporal changes brought by farmers in their farming system in the face of internal pressure (like economic necessity) and external interventions. The earlier research was conducted as a part of Master Program at Asian Institute of Technology (AIT), Bangkok. The financial support for the research was provided by Canadian International Development Agency (CIDA). The later research was conducted for the purpose of improving the previous manuscript. The author spent about two months at AIT in 1993 to improve the manuscript based on his second field study. Financial support for this purpose was provided by GTZ. Prof. Karl E. Weber provided guidance on both research studies. Accordingly, the author expresses his sincere gratitude to CIDA, GTZ and to Prof. Karl E. Weber.

A debt of gratitude is also owed to those farmers and villagers who provided answers to all sorts of questions with eagerness and enthusiasm. Since this study is concerned with increasing agricultural production and conservation of resources through farmers' participation, the author would feel successful if any of the findings of this research were used for devising programs aimed towards this end.

TABLE OF CONTENTS

LIST OF TABLES

LIST OF FIGURES

CHAPTER ONE

INTRODUCTION

This is an account of an investigation of the process of making choices generally followed by farm families of Nepal in the management of their farming operations in the light of various changes brought about by the modernization process[1] initiated by the government as well as the external agents. Changes in people's attitudes about farming and in their socio-economic structures, especially in the interrelationships among various groups, brought about by the modernization process have also been discussed.

In recent times, farmers in Nepal have been facing rapid changes in circumstances that shape their family as well as the community lives. Greater integration of the village economy with the world market forces is exposing the farmers to the changes beyond their control. Government's intervention in the form of introduction of new technology and the information and inputs associated

---

[1] Modernization here is taken as a process of social change resulting from the diffusion and adoption of the characteristics of apparently more advanced societies through societies which are apparently less advanced (Johnston et al, 1986: 302). Modernization thus represents the temporal dimension of the process of change, the movement of rural people from a 'backward' to a 'more advanced' state through the adoption of practices followed by the advanced societies. As this study deals mainly the farming operations, modernization essentially means the adoption of farming operations practiced in advanced societies of developed countries.

with that technology is also affecting farmers' behavior and their decision-making process. This intervention has also been exposing farmers to the external world. Changes in physical resources, mainly the decline in soil fertility and the availability of resources are other factors affecting farmers' decision-making process. Even though farmers in Nepal have been exposed to the external world since the opening of country to the world in 1951 after the downfall of Rana regime, rapid changes are seen since the mid-1980s. Even though this study was conducted in 1987 and 1993, it uses information about tradition of farming in the past and then discusses changes brought about by farmers in the course of time. Since household heads belonged to the older generation, information they provided on their attitudes and perception about resources, priorities and the importance of community lives pertains to the time period when they were socialized to the farming during their formative years, or rightly said, when they were living as peasants.

Decisions taken by farmers and the process utilized in taking decisions ultimately determine their adoption of new agricultural practices, which are considered important for the modernization of farming. This has been the priority of governments since the 1950s, when Nepal began to show concerns to improve people's lives. In Nepal, all development plans incepted since 1956, except the earlier two, gave top priority to the agricultural sector. In the eighth plan (1992-1997), 70 per cent of the budget had been earmarked for the rural development of which the agricultural sector is the main component. Despite such efforts, various studies (ADB, 1982; Tiwari, 1991) indicate that agricultural sector has remained stagnant even though it is still the main basis of the Nepalese economy in terms of its contribution to the employment and GDP[2]. In the 1980s and 1990s, agricultural growth rate remained less than the population growth rate, which essentially means the decline in per capita agricultural production (see, APROSC and JMA, 1995). In the current ninth plan (1998-2002), government has implemented a 20-year plan (Agriculture Perspective Plan - APP) to enhance the growth rate in agriculture production, which is assumed to bring about overall prosperity in the standard of living of the people. This Plan aims to introduce green revolution technology in a big way in order to commercialize the agriculture not only to meet the needs of domestic market, but also for export.

Because of increase in population pressure on scarce land and forest resources, the farm management and agricultural development deserve more than academic interest. A rapid and sustained increase in agricultural production is necessary to meet a growing demand for food, to raise income and to provide meaningful employment opportunities for the overwhelming large portion of the population dependent on agriculture. Development of agriculture is also needed to provide onward linkages to industries and to increase exports to

---

[2] Even though agricultural sector is said to employ over 80 % of the population of the country, its contribution to GDP has been steadily declining. Its contribution has declined to about 40 % in 1998 (Adhikari, 1999).

earn foreign exchange required for financing the importation of inputs needed for raising the economic growth. In fact, the future pace and direction of economic development currently pursued in underdeveloped countries like Nepal depend largely upon growth in agriculture and on the conservation of land and forest resources. On the other hand, only growth in production of food is not a sufficient condition to meet the food security requirements of people, especially the poor households and small farmers. It has been shown by various studies that starvation and hunger-related deaths had occurred when production was more than in normal times (Sen, 1981; Watt, 1983). In this context, agriculture development is only meaningful if food security of small farmers can be meet on a sustained basis. The significance of traditional (when farmers were peasants) and modern farming practices in meeting food security needs has been analyzed in this study. It has also been argued that the modern commercial agriculture is not conducive for food security at the household level (Shiva, 1991). Moreover, this technology is also considered as energy inefficient and environmentally unfriendly. In response to this argument, ecologically friendly farming systems have also been introduced by the conservation-related projects. The study area had also been subjected to this type of intervention. This leads to a question 'how far farmers have experienced these environmental problems and what are their perception about the changes in technology?' This study also attempts to shed light on this question.

An overwhelming proportion of Nepalese farm families can be regarded as subsistence farmers or peasants in accordance with the definition given by many authors (Wharton; 1965; Mosher, 1965)[3]. These families operate farming on a farm of less than a hectare. Farming for them is a way of life and is closely related to their family affairs. Accordingly, the decision-making process of a farmer can not be studied in isolation from his/her family and socio-cultural milieu in which he/she operates. This is especially so in villages of Nepal which are diverse ethnically and in their physical environment. Due to the existence of caste system, which is based on Hindu religion, the society is hierarchically stratified. Hindu religion and, to a lesser extent, Buddhist religion have provided ideologies that also guide peoples' behavior and influence their decision-making in farming. A family in a village of Nepal is a member of various hierarchically ordered groups like lineage and castes. These groups differ in their roles, status and occupations as ascribed by their religions and the local tradition. Moreover, the roles and status of different members of a family are also different, depending upon the age, gender and relationships within the family. Apart from the family, lineage and caste, village society is also composed of various horizontally ordered formal and informal groups, which also directly and indirectly influence the decision-making of a family. Because of the

[3] Subsistence farming is that mode of production in which the farm family's goal is self-sufficiency/ self-reliance in domestic food supply rather than production of commercial crops. The goal of production activity in cultivation, then, is family survival (Wharton, 1965). Subsistence farmers are those who sell for cash or barter less than 50 per cent of the production of their farm (Mosher, 1965).

existence of common resources like forest and pasture, which are usually managed on a customary basis, village society plays a greater role in daily lives of rural people. The physical environment in villages consists of mainly the upland, usually located at higher slopes, and lowland, usually located at lower slopes. There are also drastic variations in the temperature, in the fertility and slopes of lands and in the water availability within the boundaries of a village.

Considering the above features of village societies in Nepal, a farm household needs to take decisions to maximize or optimize its objectives - formulated taking into consideration both economic needs and cultural requirements - within the framework of opportunities and constraints provided by the diverse physical environment and the recent changes in socio-economic situations. How these physical and socio-cultural and economic environments affect the farmers of Nepal in making decisions in the production of their crops and in the management of their resources are described and analyzed in this study with the aim of examining **their rationality** in decision-making.

This study reveals, as discussed in details in the following chapters (particularly Chapters 3 and 4), that the farming system followed by farmers is very appropriate to their goals in farming, i.e., to meet food requirements of the family with least amount of external inputs. The diversified and integrated farming system was found to maximize this goal by utilizing the opportunities offered by the diverse physical environment. But total food security from farm production was not possible for most farmers. As a result they had started to look for other opportunities in non-farm areas to supplement the food produced at home. Modern technologies, especially the fertilizers and high yielding varieties, were also introduced to increase production. Owing to the decline in productivity, hard work required in farming, high prices of modern inputs and unavailability of farm laborers, farmers were found to be unsatisfied with their profession. Moreover, due to growth in non-farm employment opportunities and relatively backwardness in farming, village youths look down upon farming as an inferior profession. But in the past, farming was considered as the best profession as it provided an independent life. These findings also put question mark whether the APP can realize its goal of modernizing the farming given the situation in which youths would like to escape away from farming. Moreover, the goal of development agents to change the peasant model of farming into a commercial farming does not seem appropriate and at the same time unattainable. Studies of this type are important as they look into the perceptions of farmers and the problems they face in this transformation process.

Peasants' decision-making: theoretical consideration

As stated above peasants' decision-making, or their choice making mechanisms, can not be fully understood if socio-cultural and physical environments are not taken into consideration. Thus, the study of this type requires insights from different disciplines like sociology, anthropology, economics and agricultural sciences.

Decision-making is, in essence, a continuous process of choosing the best among alternative choices. A farmer has to take decisions more often than any other type of entrepreneurs. A producer farmer has to take decisions in many activities which are generally grouped into five broad categories viz: what to produce; what method of production to use; how much to produce; when to buy and sell; and, where to buy and sell? (Upton, 1973:3). Considering the time frame, decisions are categorized into two types viz: long-term and short-term decisions. Long-term decisions are related to location of farms, system of farming, size of the farms, purchase or renting of land, capital borrowing, investment, labor employment, and organization and orientation of the enterprise. Short-run decisions are generally related to current input use, storing and marketing.

Decisions are taken to solve problems and decisions with which farmers are confronted with relate to one or a combination of these following three questions: goals or objectives to be attained, limited amount of resources to use in reaching these goals and objectives and a number of alternative ways to use the limited resources in attempting to attain the goals and objectives (Ronald, 1981:7). It is also due to this reason that decision-making is usually defined as allocating limited resources among alternative and usually competent uses with the aim of maximizing the benefits (Dillion, 1980:257; Dillion, 1965:175; Emery, 1972:3; Heady and Jenson, 1954:6; Upton, 1976:32). This definition of the decision-making gives some indications as to the process involved in making decisions. For example, Kay (1981:9) formalized the process of decision- making into a set of logical and orderly series of steps viz: identifying and defining the problem, collecting relevant data, facts and information, identifying and analyzing alternative solutions, making decision - selecting the best alternative solution, implementation of decisions, observation of the results and bearing the responsibility for the outcome. Norton and Mumford (1982:87) considered the decision-making as a dynamic process because they found, while analyzing pest control decisions, that the decision-making is affected by the previous decisions and the experiences. In the same line, Anderson et al (1985:12) outlined the following steps for decision-making which they considered as a logical procedure for making risky choices.

1. defining relevant acts, states and their consequences;
2. eliciting prior degrees of belief or probability for the states and degrees of preferences or utilities for consequences;
3. taking account of whatever further predictive information may be available as a basis for revising the initial probabilities; and
4. selecting an optimum strategy on the basis of maximizing expected utility.

The decision-making procedures as discussed above were developed considering the farming as a business enterprise separated from family life and the socio-cultural environment in which it operates. These approaches assume the profit maximization or optimization as the sole criterion or objective of the decision-making as is generally the case in the commercial farming of the developed countries. But in developing countries, farming is subsistence oriented and farms are typically of the small type of business, in which most or all of the labor is family labor, and where the farmer himself is the sole supplier of risk capital (Ruthenberg, 1976: 2). Farming in these countries is viewed as both a production and consumption unit. Thus, the farm business is integrated with the family life and the decision-making is influenced not only by the resources the family possesses but also by the worldviews of its members and by their social organizations as these factors also influence the objective-defining-criteria and the decision-making process like collection of information, consultation and experimentation. Moreover, farming in developing countries is also integrated with common resources like forests and pastures owned and managed, in most of the cases, by the village society. These issues are not taken into consideration by quantified models of decision-making developed by the economists and thus they give us only the partial picture of decision-making behavior of the villagers. But these issues, i.e., the role of the non-economic factors and the village society on the decision- making, are taken up mostly by social scientists - sociologists and anthropologists. They generally discuss the decision-making in relation to the household and village society. For example, Firth (1965: 27) identified the following structural features that affect the decision-making process in an agrarian society: neighbor ties; family and household ties; ties of kinship; ties of ritualized friendship and of quasi-kinship such as godparenthood; status and class differences; relations of patronage; and clientship and leadership. This effect of social structure on the economic action is the important distinguishing feature of the subsistence and commercial farming. As Wharton states:

> "The most distinguishing features of subsistence or peasant farmers is the strong influence, if not dominant, of the socio-cultural considerations in the process of production and labor use and in the distribution or exchange of agricultural output. The whole economic decision-making and action complex is thus affected by non-economic forces. Such forces include more than just cultural and social obligations. They include institutional factors as well" (1965: 14).

A classic study on the agrarian economy was done first by Chayanov as early as 1926, who established a distinctive 'Theory of the Peasant Economy' According to this theory, production and consumption decisions in the household economy are interrelated. Each household works to the point where the household's subjective evaluation of marginal disutility of work equals its estimates of the marginal utility of the output gained. The ratio of producers to consumers in a family determines the minimum amount of labor time that must be expended by each producer to produce the socially acceptable minimum output for the household as a whole. Thus, this ratio determines the way in which household producers make decisions. The following inferences can be drawn from the 'Theory of the Peasant Economy':

1. the peasant economy involves an intrinsic social relation, also known as 'self-exploitation of labor';
2. the peasant economy reproduces itself through the family;
3. the peasant economy embodies a contradiction between human needs and the forces of production.

Even though criticism was raised and comments were formulated on Chayanov's theory, the central theme 'that the peasant economy can not be studied by applying profitability oriented capitalist economic principles' is still widely valid. Many of the Chayanov's arguments have been reflected in the 'substantivist' school of thought. Other forerunners of this concept are A.L. Kroeber and Robert Redfield. Kroeber just mentioned peasants in a survey volume that he published in 1948. Redfield, in 1950, expanded Kroeber's notion of 'part societies' with 'part culture'. Redfield (1956) coined the term 'great tradition' to describe the services for which peasants are dependent on the larger society. He saw the 'little community', a name which he gave to the peasant community and its 'little tradition' as different from the so called 'great tradition'. It is only after Redfield that anthropologists widened their scope of studies to include peasant societies in addition to their traditional preoccupation with studying tribal populations. Despite this emphasis for the study of peasantry, social scientists do not accept it as a separate socio-economic system comparable to feudalism or capitalism. Marxist scholars regard peasants as the awkward class of people (Shanin, 1973; Evans, 1988). Similarly, there is also controversy as to whether peasantry be called as subsistence or petty commodity producers. But in reality there is no dichotomy between these two modes of production. It is the degree of existence of these two types of production, which determines the actual type of a society. Shanin describes peasant societies considering the following basic facets of peasant life:

4. the peasant family farm as the basic unit of multidimensional social organization;
5. land husbandry as the main means of livelihood directly providing the major of part of consumption needs;
6. specific traditional culture related to the way of life of small communities;
7. the underdog position - the dominance of peasantry by outsiders. (Shanin,1971:238-262).

The first facet dealing with the peasant household as a unit of organization in production, consumption and distribution has been pointed out by several scholars. Nash summarizes that

> "The social units involved in production and consumption are households, not in combination. The economy is household organized" (Nash, 1971:161-178).

The relationship to land lies at the root of the peasant economy. The produce from the farm meets the basic consumption need and gives peasants relative independence from other producers and from the market. Firth (1965) also maintains that 'soil cultivation' is the primary means of livelihood of peasants. In addition, he includes fishermen and small traders in the peasantry. But in the context of Nepal, village forestry should also come under the peasant mode of production. The third factor is related to peasants' cultural pattern of a small village community. Due to this culture of peasantry, "peasants reach the level of self sufficiency unobtainable in the individual households" (Shanin, 1971:15). The fourth facet is related to the political economy, which emphasize the appropriation of surplus by outsiders who develop relationships with the elite within the peasant communities.

The inter-personal or social relationships existing in peasant societies also play a decisive role not only in their social and ritual behavior but also in their economic matters. Wolf (1977:7-10) has coined the term 'ceremonial fund' to include the expenditure incurred by peasants to organize the culturally defined ceremonies to justify and regulate various social relationships. The types and the nature of these social institutions also vary from one society to another. As wolf states further:

> "Peasants' world is not amorphous, but an ordered world, possessed of its particular organization. Moreover, these forms of organization vary from peasantry to peasantry. No one easy formula will do for all" (1966:viii).

The interpersonal relationships in peasant societies not only include relationship between peasants, but also between peasants and their superiors living within their village or outside the village. The latter form of relationship is called patron-client relationship, which is common in South Asia mainly because of the hierarchical stratification of societies due to the existence of caste system. Wolf (1966) regards these relationships as "many stranded, dyadic hierarchical coalitions". These relations are developed by the peasants' expectation of assistance from superiors in coping with the perceived outside threats. Three typical characteristics that Powell (1970) recognized in such relations are: friendships between unequal parties, reciprocity in exchange of non-comparable goods and services, and a face to face contact. This patron-client network is very distinct from the formal organizations. They operate without legally forcible sanctions. Mayer described such groups as 'quasi-groups' or 'action-sets' (Powell, 1970). The most common form of patron-client relationships in the context of South and Southeast Asia is the landlord-tenant relationship. The goods and services provided to landlords by tenants for the use of land include more concrete benefits. Due to commercialization, the terms of trade shifted in the favor of landlords, who naturally moved to take advantage of the change. As a consequence, patron-client alliances have declined.

Peasants - material or non-material maximizers: issue of rationality

Because of the role of non-economic factors on the economic behavior of the peasants or subsistence farmers, questions have often been raised on the rationality of their decision-making. Many prominent scholars in the past doubted that the economic concept of rational behavior developed in industrial economies could be of value in understanding peasants' economic behavior. Polanyi (1957:243) was of the view that peasant societies use reciprocity and redistribution rather than market forms to supply members of the community with 'material want satisfaction'. Further, he argued that 'rational choice concerning means-to-ends' need not be present under such conditions. Followers of this doctrine, including Dalton and Nash, deny the existence of decision-making and economizing behavior in such societies. Nash stresses the limits imposed by social structures on such behavior (cited in Calavan, 1977). Rogers (1965:115) noted ten elements that were generally linked with the subculture of the peasantry: mutual distrust in interpersonal relationships, lack of innovativeness, fatalism, low aspiration level, lack of deferred gratification, limited time perspective, familism, dependency upon government, localiteness and a lack of empathy.

Foster's 'image of limited goods' (quoted in Calavan, 1977:134) is related to the farmers' irrationality in decision-making. This concept explains that peasants perceive of the supply of goods, both material and non-material, in their society, as limited which causes distrust in interpersonal relationships, the consequences being the economic underdevelopment. It tells the peasants that the total volume of desirable goods in a society is fixed, so that one can increase one's share only at the expense of others, which becomes a cause of interpersonal distrust. The co-operative approach, however, adopted by peasants to increase the total value of goods can not be explained through this theory. For example, there are many cases of co-operation in village Nepal for the conservation of common resources, which has certainly been helping in slowing down the degradation of the environment.

Despite the above discussed arguments against the rationality of the peasants' economic behavior, it is unrealistic to say that they do not make decisions or that their decisions are irrational. They take action in the expectation that it brings the greatest satisfaction to them. Here, the basic problem lies in the matter of preference rating. Preference of a decision-maker or actor, in this case the farmer, should be taken into account while developing farmers' decision models. In this context, social scientists use the term utility (preference) to distinguish from profit or output, because the relationship of preference and outcome is not as direct in the case of a peasant as it would be in the case of a commercial farmer. For a peasant, the significance of an outcome does not depend on physical production but on the number of wants it may satisfy (Ortiz, 1967). Complexity arises, in the case of the peasant economy, in defining wants because these are socially defined due to multistranded relationships and close nature of their societies, which may vary from culture to culture. The produce is evaluated in terms of its satisfying power to meet consumption demands, which are socially and physically required by a producer. Thus the concept of rationality should be looked at not as a simple means and end relation but as a sequence of social actions (Ortiz, 1967).

It has also been discussed that cultural norms play an important role in decision-making which, in some cases, may seem to lead farmers towards irrationality. But Moerman (1968) argues that this is not so. Maximizing strategies are also contained in the culture, which guides farmers to follow traditional practices. Experiences about appropriate means and ways of production of the past generations are cumulated which become a culture or tradition and it contains 'rules of thumb' for guiding farmers. Moerman (1968) called them a 'grammar of farming'. Thus, following a tradition does not mean that it is an irrational behavior. As Wolf states:

> "To say that a society is traditional or that its population is bound by tradition, does not explain why tradition persists, nor why people cleave to it. Persistence like change is not a cause - it is an effect" (1966: viii).

It has also been argued that various traditional farming systems are best suited in the given physical environment. Lipton (1968:334) identified some of these traditional practices that are truly optimizing in the case of tropical agriculture, e.g., covering the manure pit, ploughing across the slope, producing more output for least input, higher seed rate in rainfed areas and mixing legume with non-legume crops. Moreover, it is also unlikely that peasant take the tradition as the rigid rule. They re-evaluate the practices in the light of changing circumstances and adopt that art of cultivation, which does not negatively influence their strategy of farming. On the comment about the slow response to new circumstances, Ortiz (1968) argues that constant re-evaluation of means in terms of preferences can be extremely wasteful. It requires time and increases uncertainty. Moreover, the tradition, habit, impulse and apparently irrational behavior of conspicuous giving and destruction (eg. *kula* or *potlatch*) also serve as income leveling device and as a mechanism for social security. For example, distribution of meat by a lucky person is a way of ensuring protein intake. Peasant interest in having many children, even at the time of food scarcity, is due to their expectation of benefits in their old age (Fallon, 1983). Similarly, ceremonial rewards and sanctions reinforce social and economic relationships and serve to limit destructive behavior. Demaine (1978: 49-67) shows the significance of such ceremonial and social sanctions in the effective control and regulation of rain water distribution in the context of traditional irrigation system of South East Asia. He even suggests to attach ritual significance to the organization of present irrigation water supplies for effective distribution.

Many contemporary scholars believe that farmers are materialistic maximizers. Schultz (1964: 37) is an economist who has strongly emphasized the role of the farmer as economic man. He tested and proved the hypothesis, which says that farmers in developing countries are efficient but poor.

Risk and uncertainties in farming

One of the reasons to call peasants or subsistence farmers as irrational is because of their risk-averse attitude due to which it is said that they are hesitant to adopt modern technology. Binswanger (1978) points to the fact that risk aversion can arise from farmers' attitude towards risk, i.e., willingness to take risk and from their inability to take risk because of the lack of resources or security. Moreover a peasant farmer is especially prone to risk and uncertainties because he usually does not possess either the information or the technical knowledge to make accurate forecasts of yields, of pay-offs or natural conditions, which leads to an uncertain situation. Dasgupta (1970:36-61) has classified uncertainties in farming into three broad categories:

1. Limited knowledge of the state of nature which includes, in turn:
   a. uncertainties arising from the role played by natural inputs
   b. uncertainties arising from variations in labor requirements due to yield variations; and
   c. uncertainties arising from market price fluctuations.
2. New technological possibilities.
3. Changes affecting the social framework in which an individual operates, e.g., land reform.

Because of uncontrollable natural factors like climatic condition, light, heat, rainfall or soil condition, farmers in developing countries have to face numerous uncertainties. Variations in the impacts of these factors cause farmers to choose different types of crops. Regarding the changes in prices, many economists believe that subsistence farmers do not respond to such changes. But it has also been revealed that changes in price at least affect the cash crop component, which subsistence farmers grow in anticipation of some cash income from the sale of such crops in order to buy some consumer items from the market. It may also be argued that price fluctuations indirectly influence the peasant-farming pattern due to the influence of some few surplus producers who show positive response on such situation. The programs like land reform, land right legislation and tenancy system, among others, also cause uncertainties for such programs have been a matter of discussion in policy making forums, and in some countries they are still in the process of implementation.

To cope with the uncertainties about the future outcome resulted from the above discussed factors, farmers subjectively evaluate the range of possible kinds of outcome. Past experiences play an important role in such forecasting. Farmers also develop a tradition or culture of forecasting the situation. It has also been said that farmers, being risk averse, do not adopt any improved technology. This argument is contradicted when we say, firstly, that all the farmers are taking decisions in their daily business, and secondly, that there are instances of adoption of new technology or investment in new activities when such investments do not affect their household strategy.

Moral or political economy : relevance to decision-making

The issue of rationality of farmers' economic behavior including risk taking can be analyzed from the level of farm household or from the level of society. Political economists study the issue of rationality from the point of view of individuals or their families. But moral economists' view of farmers' decision-making is focussed on collective benefits. They maintain that peasants are risk averse and have, therefore, collectively developed social insurance mechanisms. Following paragraph drawn from Scott (1977) illustrates this point:

> "The precapitalist community was, in a sense, organized around this problem of minimum income - organized to minimize the risk to which its members were exposed by virtue of its limited techniques and the caprice of nature. Traditional form of patron-client relationships, reciprocity and redistributive mechanisms may be seen from this perspective" (1976:9).

In contrast to moral economists' view of farmers' preference for risk aversion, Popkin states that :

> "Although poor and close to the margin, however, there are still many occasions when peasants do have some surplus and do make risky investment: the fact that they are poor and risk averse does not imply, either logically or factually, that they do not make investments" (1977:18).

As stated above, the relevancy of this debate lies in the approach of the study, i.e., whether rationality be looked into from society as a whole or from an individual farmer's point of view. The study of rationality from the individual household level and from the system level may lead to different conclusions as Popkin (1977:31) says that 'what is rational for an individual may be very different from what is rational for an entire village or collective'. One suitable example is fertility. More household labor force at the time of need is rational at the household level, whereas the same consideration may cause a population problem at the system level. The 'tragedy of commons' is another problem resulted from the habit of maximizing individual benefits at the cost of social benefits. Therefore, the study of rationality in farmers' behavior also needs to consider the actions of individual farmers or families in the light of the consequences of these actions at the level of the society.

Farmers in less developed society are thus caught in a dilemma: whether to maximize individual or household-centered benefit (rationality) or follow the moral principles defined by the society so that community welfare is not sacrificed. As seen in this study, farmers were very much obliged by the moral economy in the past. But in recent times, farmers were found to have given emphasis to household-centered rationality. This shift in farmers' attitudes has occurred mainly due to greater integration of village economy with the market and the growing economic necessity to look for other sources of income.

B. The framework of analysis and organization of the study

The above discussion provides us the framework for the analysis and description of the farmers' decision-making. The main issues of this framework are: peasants decision-making should not be confined only at the household level but also include decision-making at the village level; physical environment and socio-cultural features should be taken into account while studying the goals and objectives of farmers and their decision-making process like collection of information, consultation, choosing alternatives, implementation and evaluation; role of experience gained from the action of the past, i.e., implementation and evaluation of earlier decisions, for the decision-making in future. This framework is illustrated in Fig. 1.1.

The organization of this research work is also based on the framework of analysis discussed above. Chapter 2 discusses the physical, cultural and socio-economic features of the study area under which farming as a way of life has been carried out. Chapter 3 discusses the role of ecological factors in shaping the farming system of the study area. Especially this chapter describes how farmers have utilized various resources available in their villages in fulfilling their objectives in farming. Chapter 4 describes the process and technique of decision-making at the family and village levels. The role of socio-psychological and socio-cultural environments in the process and technique of

Fig. 1.1: Framework for the analysis and description of peasants' decision-making on the production of crops and on the management of resources including the common resources.

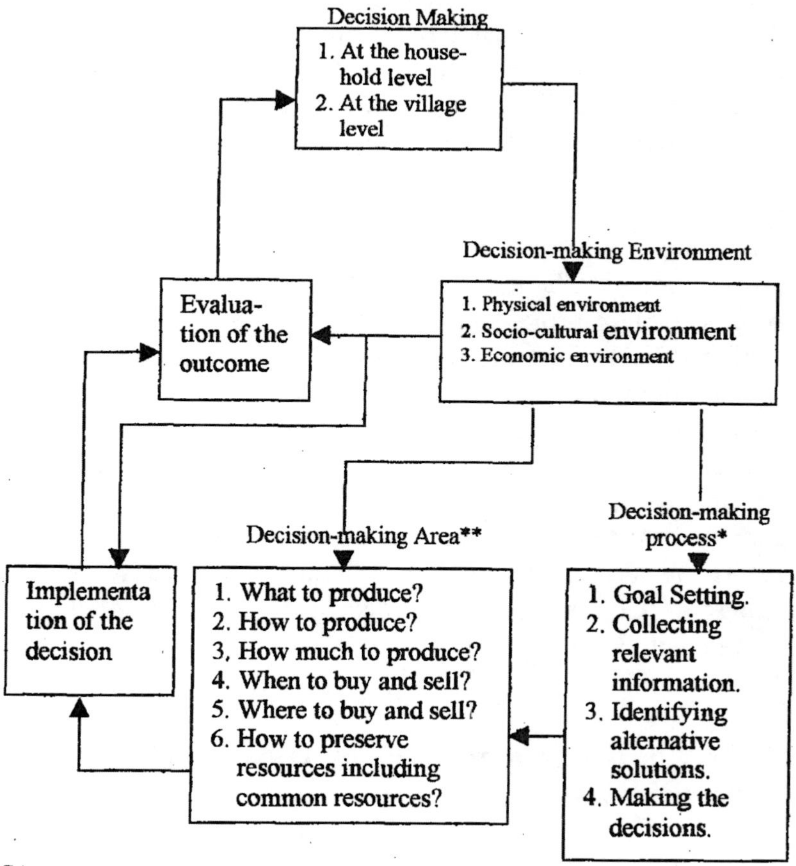

Sources:
* Adopted from Kay (1981:9) ** Adopted from Upton (1973:3) – Last item was added by the author.

decision-making is described in this Chapter. In essence, this Chapter examines how the decisions in the use of resources as discussed in Chapter 3 were made. Farmers' perception about the farming life, their goals in farming, their problems and their beliefs on astro-magical and other ritual activities are also included in this Chapter. Chapter 5 integrates the findings of Chapters 3 and 4 to examine the main proposition of this study, i.e., the rationality of farmers' decision-making. Important quantitative and qualitative variables related to physical, cultural and socio-economic environments were identified from the discussions in Chapters 3 and 4 and their effects on the adoption of improved technology of rice production was analyzed using a quantitative model (QUANT II) capable of including quantitative and qualitative variables. The main purpose of this statistical analysis was to examine whether socio-cultural environments have been inhibiting farmers to adopt modern farming practices. Impressions gained from the re-visit to the study area after 6 years (in 1993) are also included to elucidate the flexibility of the villagers. Conclusions and recommendations for planning are presented in Chapter 6.

C. Development context: decision-making and peoples' participation

The study of decision-making at the household and village levels is clearly important from the point of view of motivating villagers to involve or participate in developmental projects. Studies of decision-making are helpful in this regard because they reveal the constraints and opportunities provided by socio-economic and cultural institutions for different income and cultural groups to participate in developmental programs. In essence, it helps us to understand the willingness and ability of different categories of farmers to participate. For example, various studies have shown the unwillingness of poor households to adopt cash crops or trees in their farms because of their inability to take risks resulted by price fluctuations in the former case and by the long gestation period of income in the latter case (Hudson, 1980:52; Lado, 1988:57). But high income group had participated more actively in such projects. On the other hand, it is also reported in the context of Philippines that soil conservation measures were effectively adopted by small farmers because it directly helped them to secure their subsistence need more safely (Segura-de los Angeles, 1986:179). This finding tells us that people participate in developmental works if it matches with the peoples' goals and objectives and their resource possession. Poudyal's (1990) study conducted in Nepal also leads to similar conclusion as he found the environment for participation favorable if needs and expected benefits from an activity, capacity of people to participate, and possible scope of participation coincide. Similarly, existence or non-existence of social relationships and of the system of group decision-making also affect the participation. For example, it is seen in the districts of Kerala, India, where farmers and laborers were organized for a longer period into associations or unions derived more income and the caste differences in social status had been narrowed (Uphoff, 1978). Uphoff (1978) further concluded that this organizational factor compensated for the inhibitions raised by factors like age, sex, race, education, caste, class or land tenure status. The study of decision-making is thus important to understand what motivates various categories of farmers to participate in any activity.

Poverty alleviation has been the focus of the many developing countries partly because of the emphasis on human development and partly because of the need to conserve and maintain the environmental balance. This is also the case in Nepal : the ninth plan (1998-2002) has emphasized the poverty alleviation as the main priority area. The main strategy has been to bring poor people in the mainstream of development. People's participation in developmental work means that they need to be involved in the planning, decision-making, implementation and evaluation of such work. Planning essentially means an organized, conscious and continual attempt to select the best available alternatives to achieve specific goals (Waterson, 1965:26). Since the need and priority of different groups of people within a village community is different, problem of conflict may often arise. Similarly, mode of participation of people is also different depending upon their resources, income, history of the society and social and family organizations. Therefore, development workers including policy makers, planners as well as extension workers working at the grass-roots level should have the complete understanding of the working (decision-making) of the village families and their communities.

In recent times, the rural youths have shown a tendency to move away from farming. Others engaged in farming have also beginning to show dissatisfaction over farming as a profession. In these circumstances, it is necessary to understand the perception, attitudes and aspirations of farmers. Unless these aspects of the farmers' lives are not understood, it is difficult to bring desired changes in farming

D. Research design:

This research is based on a case study of Begnas Village Panchayat[4], hereafter called Begnas Village Development Committee (VDC), of Kaski District, Nepal. Information was collected from both the primary and secondary sources. Secondary data were required to describe locational, climatic, physical and other related aspects of the study area. The main part of the study is based on the primary data that was collected by interview method using the structured questionnaire.

---

[4] Village Panchayat was the lowest administrative unit during the partyless Panchayat Political System which was in vogue from 1960 to 1990 in Nepal. A Village Panchayat consisted nine wards. After the restoration of multiparty political system in 1990, Village Panchayats were called Village Development Committees (VDCs). The village boundaries of this Village Panchayat were also redefined. Now, Ward 9 of the former Village Panchayat has been excluded. Since this study was conducted in 1987 covering the whole Village Panchayat, the Begnas VDC here refers to the geographical area of that Panchayat.

Secondary data were collected from sources like VDC offices, schools, government census reports and other official records. Observation and information collected from key informants were also used for the gathering of certain facts, which were found to be difficult to collect employing a structured questionnaire. The author - born and raised in an average farm family in a village near Pokhara - also used his experiences to describe the village life, particularly the influence of socio-cultural environment on farm activities.

Questions related to the perceptions of farmers to different decision-making environments were asked after telling a short story related to a farmer who takes decisions under certain hypothetical conditions. Then, the interviewees were asked what decisions they would have taken in that situation, and whether such situations have actually occurred in their practical lives. Singh and Ray (1980) used a similar kind of approach to study farmers' aspiration level. They framed a story in Hindi and recited it to farmers before asking questions.

Sampling and sample size

The sample size for the questionnaire survey was determined using 'Tables for Statisticians' prepared by Arkin and Colton (1963). Taking 95 per cent confidence level and 4 per cent reliability sample size for the population of the study area, which was more than 1,000 was estimated at 266. Simple Random Sampling method was used to identify the sample households. After obtaining the population ward-wise, numbers starting from 0001 were given, corresponding to the population size of the village. Then a simple random number was drawn from SRS number table, which was later assigned to the concerned ward. Twenty per cent, i.e., 53 more numbers were drawn and assigned to the wards in the same way for reserve. Thus, in total, a sample of 319 respondents was drawn. Reserve sample was utilized on a first come first serve basis. Only 16 reserve sample households were utilized. The questionnaire for the household survey was first prepared and then pretested in a village near Pokhara, Nepal. After the necessary revision, the questionnaire was translated into Nepali and then printed.

Data analysis

Various descriptive as well as analytical statistical methods have been used to analyze the data and draw conclusions. Descriptive statistics like arithmetic mean, standard deviation, range and correlation were used. Quantification Theory II was used to analyze the factors affecting the farmers' decisions on the adoption of modern technology in rice cultivation. For this purpose, Quant 2 package was used. Quantification Theory II facilitated the inclusion of various socio-cultural variables, which otherwise were difficult to quantify into the model. Following indices have been constructed for the quantitative interpretation of qualitative data, and these indices simplified the analysis and interpretation, and facilitated the comparison.

a. Index of satisfaction:

The purpose of this index was to find out the overall satisfaction level of farmers and to compare their satisfaction with non-economic and economic attributes of farming. For this purpose, the following type of index was used.

ISA = [FS(1.0) + FaS(0.6) + LS(0.2) + NA(0.5)]/N
IST = (Sum of ISA)/Na

where,

ISA = Index of satisfaction with individual attributes
FS = Fully satisfied
FaS = Fairly satisfied
LS = Less satisfied
NA = Not available
N = Frequency
IST = Index of satisfaction from total attributes
Na = Total number of attributes.

Twelve attributes were given for scaling, six each from financial and non-financial sets.

b. Index of farmers' perceived problems:

The purpose of this index was to measure the relative severity of the problems to farmers and for this the following index was used.

IIP = [IP(1.0) + FIP(0.6) + LIP(o.2) + NIP(0.0) + NA(0.5)]/N
IGP = [Sum of IIP (of group)]/Ni

ITP = [Sum IIP]/Na

where,

IIP = Index of individual problem
IP = Important problem
FIP = Fairly important problem
LIP = Less important problem
NIP = Not important problem
NA = Not available
N = Frequency
IGP = Index of group problem
Ni = Number of problems in a group
ITP = Index of total problems
Na = Number of total problems

Twenty-two problems related to land, labor, inputs, credit and marketing were given to farmers, to scale on the basis of the severity of the problems to them.

b. Index of fatalistic attitude and magico-astrological beliefs:

Two separate indices were prepared to measure farmers' fatalistic attitudes and magico-astrological beliefs.

IIB = [SB(1.0) + FB (0.6) + WB (0.2) + NA (0.5)]/N
TI = (Sum of IIB)/Na

where,

IIB = Index of Individual belief
SB = Strong belief
TI = Total index
Na = Number of belief subjects
FB = Fair belief
WB = Weak belief
NA = Not available
N = Frequency.

Five statements reflecting the commonly held fatalistic beliefs and six statements related to magico-astrological beliefs were given to farmers to scale based on the degree of their faith on these statements.

d. Index of awareness of modern technology:

Index to measure farmers' awareness of modern technology was constructed as follows.

IAI = [FA(1.0) + SA(0.5) + NA(0.0)]/N
IAT = (Sum IAI)/Na

where,

IAI = Index of awareness of individual technologies
FA = Fully aware
SA = Slightly aware
NA = Not aware
N = Frequency
Na = Total number of technologies

Five types of technology related to crop production and appropriate to the study area were given to respondents for scaling their awareness.

An effort has been made to analyze the data by education and farm size. In some cases, analysis is also done from locational point of view. Composite family education was calculated by giving different weight to different educational levels (0 for illiterate, 1 for primary school or literate, 2 for secondary school, 3 for high school, and 4 for college and above). On this basis, Total Educational Score (ES) for a family was calculated which was then divided by number of school age family members.

Then, families were classified into low (ES less than 1.0), medium (ES more than or equal to 1.0 and less than or equal to 2.0) and high (ES more than 2.0) educational levels. Basis for this was mean educational score of a family, which was slightly more than 1.0. Taking the average land holding into consideration, families were also classified into landless (less than five *ropani*), small (more than or equal to five *ropani* and less than 15 *ropani*), medium (more than or equal to 15 *ropani* and less than 30 *ropani*) and large (more than or equal to 30 *ropani*). Even though the importance of non-farm income has grown significantly in villages of Nepal, land is still the main source of income. It is thus assumed here that land holding of a household represents its economic status. Moreover, until recently, land ownership was considered as one of the status symbols in villages. As a result households used to invest their savings mainly in land.

CHAPTER TWO

THE VILLAGE

ITS PHYSICAL, CULTURAL AND SOCIO-ECONOMIC SETTING

Field research for this study was carried out at Begnas Village Development Committee (VDC) during the period between May and June, 1987. This study area was selected on the basis of the researchers' familiarity with the location. In this chapter, the physical, cultural and socio-economic setting of the village is described, within which farming as a way of life is being undertaken by its population.

A. The physical setting

Begnas VDC lies about fifteen kilometers east of Pokhara Town, the headquarters of Kaski District. Pokhara is also the headquarters of the Western Development Region (Figs. 2.1 and 2.2). An unpaved road branches off the Pokhara-Kathmandu highway and extends up to the village. A local bus from Pokhara Town provides transportation facilities for the people. This road has also facilitated the flow of goods and service. The altitude of the village ranges between 600 and 1500 meter above sea level. The village is located on the mountain flank with southwestern exposure. A small part of the village is located on the hill flank with eastern exposure. The northeastern slope of the mountain is

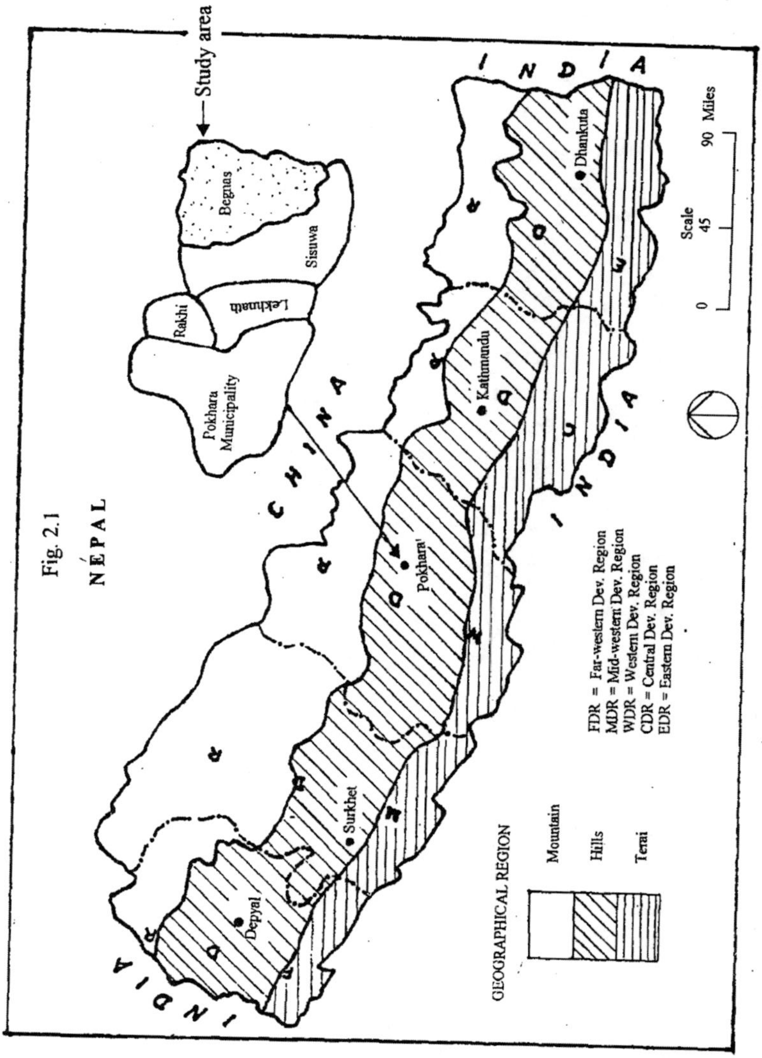
Study area
Begnas
Sisuwa
Lekhnath
Rakhi
Pokhara Municipality
CHINA
INDIA
Pokhara
Kathmandu
Dhankuta
Surkhet
Depyal
GEOGRAPHICAL REGION
Mountain
Hills
Terai
FDR = Far-western Dev. Region
MDR = Mid-western Dev. Region
WDR = Western Dev. Region
CDR = Central Dev. Region
EDR = Eastern Dev. Region
Scale
0
45
90 Miles

Fig. 2.1

NÉPAL

Fig 2.2: The study area [Begnas Village Development Committee (VDC)*]

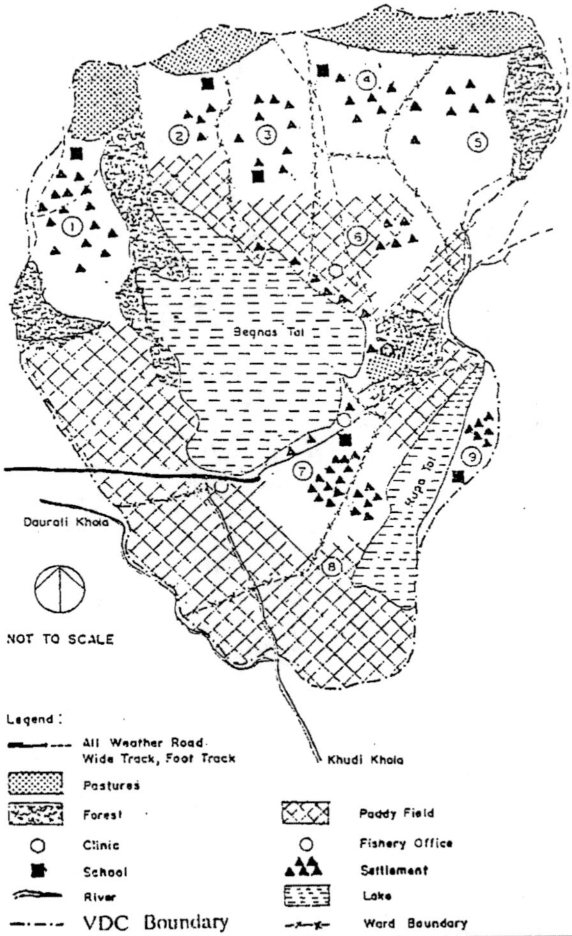

* Also includes Ward 9 of the former Begnas Village Panchayat. Ward division in this map was made according to the situation that existed in 1987 when the fieldwork for this research was conducted.

very steep and covered by forests. Thus, the village is largely a hilly village with some plain paddy fields at the bottom of the mountain. Two small lakes, Begnas Tal and Rupa Tal, are located at the foot of the hill.

According to the data available in 1987 from the then Panchayat office, the total area of the village was 2,644 ha, of which 21.2 per cent was the agricultural land, 44.8 per cent forest, 5.8 per cent pasture, 13.8 per cent residential area and the rest covered by water. But the data obtained from the Begnas Tal Rupa Tal (BTRT) Project showed that 52.4 per cent of the area was used for farming and 29.4 per cent covered by the forest. Since the data obtained from BTRT Project was based on the survey conducted by the Project in 1985, they were considered more reliable. With the population of 5,318 persons, population density on total cultivated area was estimated at 9.5 persons per ha, whereas the same figure for Nepal and Kaski District was 5.7 and 5.5 respectively. Thus, the population density of the study area was very high and this study can throw some light on how the people in such a densely populated area have been using their local resources to make a living. Even though the population of this VDC according to 1991 census was 6,803, the changes in geographical boundaries and inter-village migration made it difficult to analyze the population growth.

Climate:

The climate in Begnas was of the humid subtropical type marked by a monsoon rainfall pattern. It received about 3,500 mm rainfall, which is concentrated mainly in four summer months (May, June, July and August – see Fig. 3.1). Farmers also faced inter-year variations in rainfall, which was a cause for substantial doubt and anxiety among the farmers. In some years, very heavy rainfall could damage crops and intensify erosion. But there was less variation from year to year as to when rainfall starts to fall with regard to precipitation below 200 mm per month, particularly in November. Thus, there is not much problem for farmers to predict quite accurately when their fields will start getting dry. Uncertainty of rainfall is a problem not only for the production of rainfed rice but also of irrigated rice, since such lands are irrigated by water from springs, whose capacity depends upon initial rainfall.

Regarding temperature, there was slight year to year variation. Thus, temperature fluctuation is not a matter of concern for the villagers. Indeed, freezing weather, which occurred during the months of December through February, could damage such crops as potato. Hail damage was a serious problem for the farmers in the area. It generally occurred in two periods of the years:

1. During September through November when the paddy crop can be considerably damaged; and
2. During February through June when the maize crop can be damaged.

Temperatures within the village boundaries varied from place to place. Rice farming was done at the foothill or at the bottom of hill, because of high temperature at such places. It was common in all hilly villages of Nepal. The temperature at the top of the hill was lower, and this location was occupied by residences. Other reasons for concentrating residences at the top of the hill were the availability of fresh air fanned by a breeze and the lower occurrence of insects, particularly mosquitoes.

Vegetation:

Because of the diverse environment, different types of fauna and flora were found to exist in the study area. Deciduous trees were native to and typical of this area. Some important trees were *katus* (*Castranopsis indica*), *chilaune* (*Schima wallichi*), *sal* (*Shorea robusta*), *sindhure* (*Mallotus phillipensis*) and *tiju* (*Dodecania spp.*). Trees like *Engelharodita*, *Rhyzigium*, *Myrica*, *Rhus*, *Quercus*, *Betula*, *Cedrela*, *Lagerstoemi*, *Anthocephalus*, *Albizzia*, *Rhododendron*, *Eurya* and *Prunus* were also available in the local forests. *Imperata* and *Artimisia* were the main grass species. Farmers also grew different types of fodder trees suitable for the various types of livestock. Poisonous as well as medicinal herbs for both human and animal consumption were also growing wild. Farmers had knowledge of such trees, shrubs and herbs as well as of their characteristics.

Owing to the variation in temperature within the village boundaries, many types of fruits of both tropical as well as temperate types were grown. Tropical fruits included mango, jack-fruit, banana and pineapple. Temperate fruits included pear, plum, peaches and citrus such as lemon, orange, and mandarin. In most cases, fodder and fruit trees were not grown on separate pieces of land but were planted on the levees of terraces or at the edge of fields.

Soil:

Soil types and qualities differed between the lower plain area and that at higher altitude. Soils at the foot-hill were of the clay type and rich in humus, while that at the top of the mountain were of the silty-loam type and poor in organic matter. Though the scientific nutritional status of soils was not known to villagers, they classified land into four types depending upon soil quality and/or water availability. Paddy land was classified considering both parameters mentioned above, whereas upland was classified solely on the basis of soil quality. Soil quality was gauged by the trend of past yields. The four classes of soil were: *abal* (very good), *doyam* (good), *sim* (medium) and *char* (poor). This classification was done from a locational point of view rather than considering individual plots. Thus, certain pockets were considered *abal*, or *doyam*, or *sim*, or *char* without recognizing certain minor variations within a specific location. But when members of the cadastral survey team measured the land in 1975, they labelled each plot of an owner according to the quality for tax purpose. Crop cultivation, land-rent or tenant share and sale prices were still based on this traditional soil classification.

Villagers also had a tradition of evaluating paddy land by the criterion of quality of water that irrigates the land. Those lands which received cold and fresh water were considered best because, as villagers believe, best quality rice could be grown in such land. In the context of this village, which lacked irrigation facilities, the distinction between rainfed and irrigated paddy fields needed clarification. Rainfed were those fields on which transplanting of paddy was done when there was accumulation of sufficient water from rain for the puddling of soil. After that only rain water irrigated the crop. Locally, these fields were called ***bhal chopuwa***. In the case of irrigated fields, irrigation was done using the water from springs, which sprout only after the onset of the monsoon rains. Once sprouted, they provided water till the start of the winter season. Different practices for the cultivation of paddy were adopted in the rainfed and irrigated fields in order to cope with the availability or scarcity of water. This issue has been taken up in the next chapter.

Settlement pattern:

Clusters of houses were distributed all over the study area. The degree of the compactness of each cluster was also found to vary with the cultural patterns of the residents, location in relation to the market and the time when houses were made. People belonging to the Tibeto-Burman group like Gurungs and Magars generally built their houses compactly which was mainly because of their community way of life. Similarly houses close to the market were compactly built. Settlements established in the past were more compact in comparison to those established recently.

Various clusters of the houses in the study area could be grouped into four aggregations, even though administratively it was divided into nine wards.

Functionally, these nine wards could be grouped into the above mentioned four aggregations or subsettlements. The oldest subsettlement was Begnas, which lied almost at the center of the Village, but at the highest altitude. This subsettlement covered Wards 2, 3, 4 and 6 as well as some part of Ward 5. Wards 1 and 9 each formed a subsettlement, called Lamaswara and Bhangara, respectively, whereas Wards 7 and 8 with the remaining part of Ward 5 formed yet another subsettlement known as Pachabaiya. All these subsettlements were located at the top of the mountain range, surrounded by upland called *bari*[5] and by jungle.

Pachabaiya was located at the mouth of the village and was linked to Pokhara Town by a motorable road. This subsettlement showed a distinct feature in which it differed from the other subsettlements. It was more compact than any other subsettlement and was growing fast. It was the source of supply of necessary goods and services not only to this village but also for other adjacent villages. The people of Begnas had to walk about three to four hours to come down to the point from where vehicles were available to go to Pokhara Town. For this purpose, people of Bhangara had to walk two to three hours, whereas those of Lamaswara had to walk one and half hours.

[5] Generally *bari* and *pakho* was used interchangeably. But some slight differences were also found in the use of these terms. *Bari* specifically denoted well-manured and regularly cultivated dry land. *Pakho* represented a tract of less manured land along with the uncultivated surrounding. People also used a term '*pakho-bari*' to indicate both types of land.

The settlement pattern was more of the cluster type at Begnas and less so at Lamaswara and at Bhangara. But the houses of Gurungs in Bhangra were compactly built than in any other subsettlements. The scattered clusters of houses in Pachabaiya was the result of the immigration of Thakalis and Newars, most of whom had settled there to run shops. These business-oriented people started building houses along the roads, and as a result the once a close settlement became dispersed.

Houses in the study area were generally built of stone staked with mud mortar and roofed by stone slate, thatch grass or corrugated zinc sheets. In the past, houses were roofed either by stone slate or by thatch grass, the former in the case of rich families and the latter in the case of poor families. But due to the scarcity of these roofing materials, more and more houses in recent times were roofed by corrugated zinc sheets. Many of the poor families were still not able to afford the zinc sheets and their houses were still being roofed by thatch or by other vegetative matters. Therefore, the housing conditions easily reflected the economic position of the households.

Location of Cropping land in relation to settlements:

The existing settlement pattern has been a long established one, although new houses are appearing here and there. The physical location of the house in relation to *bari* (upland) and *khet* (lowland) and to forest showed an interesting pattern. Houses were located, in most of the cases, in the center of the *bari* which were generally situated at the upper part of the hill. Pasture was also near these subsettlements. Livestock being a part of farming in the village, the location of subsettlements near forest and pasture was inevitable. This locational arrangement would certainly help to reduce the time required for the collection of fodder, leaf litter, firewood and timber from the forest and for taking animals to the pasture and then to home. Most houses had their *bari* in one or two locations, but generally these were adjacent to the houses. Similarly, grassfields generally called as *kharbari* were also located in the upland close to the household.

Khet land was generally located at a distant place from the house, because paddy lands were generally found at the bottom of the hill whereas residences were at the top. The relative distance to *bari* and *khet* was related to different manuring, land tenure and other practices which also influenced the social life of the villagers. Different types of crop grown in *bari* and *khet* were largely determined by the ecological factor. Indeed, there also were differences in varieties of crops grown at different locations within the *bari* and *khet* areas. Paddy fields of one contiguous location were called *phat*. 'Syankhudi', 'Dhode', 'Sera', 'Talbasi', 'Gairi' and 'Libdi Maladi' were the *phats* within the village boundaries. These *phats* received water, and wet paddy was grown in these areas. Rainfed paddy fields were located at the bottom of the hill just above these *phats*. Farmers of all four subsettlements held these three types of land, i.e., upland, irrigated and rainfed *khet*, located at different places. Most of the farmers of Begnas had their land in Gairi and Libdi Maladi, most of which was rainfed. Farmers of Lamaswara had their land in Dhode and Synkhudi. Sera and Talbasi land was mostly owned by farmers of Pachabaiya and Bhangara. But some farmers had land located in all these places.

Water and other natural resources:

Water and other natural resources were available in the study area, but their utilization was at a minimum. There were three small but perennial rivers. Two lakes, which occupied 35.7 ha. of the area, were also largely unutilized. About thirty *pode* (fisherman caste) families, which occupied the lowest position in the caste hierarchy and whose traditional occupation was fishing were using the lakes to raise fish in cages. Children of these families collected a type of fruit (*simal kande*) grown in the lake which provided them some foods for snacks. But lately members of higher castes were also participating in fishing business as a source of cash income. Water was very scarce in Begnas and Bhangara subsettlements and thus cash crop growing was seriously hampered there. On the other hand, taming the river for irrigation was also difficult, owing to their more or less vertical flow. The Synkhudi river formed an inlet to Begnas Lake, whereas the Khudi river formed its outlet. The Khudi river was utilized for irrigation by the farmers of other villages. The newly constructed dam in the Begnas lake and the new irrigation infrastructure had facilitated the flow of water to low-lying villages. The Talbasi river was an inlet-river of Rupa Lake. This river was utilized to some extent for irrigation in Talbasi *phat*. Farmers had dug irrigation channels and built small dams or levees to direct water into their fields. The other source of irrigation was water from springs, but these were solely seasonal. Farmers had dug channels to bring spring water to different fields. Maintenance of such channels was a collective responsibility. Before the start of the rainy season, all farmers would come together to dredge and scrape the base of a channel, and repair the levees (of terraces) where necessary. The system of irrigation was by natural gravity flow. Terraces also facilitated directing water flow from one location to another.

B. The cultural setting

The cultural setting of any area is intimately related to the history of not only that particular area but also to that of the nation-state. The history of the Begnas VDC was also found to have been affected by that of the formation of Nepal as a nation-state, which ultimately affected its cultural setting. Even though, there were no records and documents from which the history of this study area could be reconstructed, remnants of the past buildings and the local folklore provided some sources of information.

It appears that Begnas along with other adjoining villages was one of the micro-states then existed before the unification of the country which started in 1747 and completed in 1816 after a peace treaty with British-India. The kings or chieftains were, in all probability, the Magars, one of the original settlers belonging to the Tibeto-Burman group. Migration of members of Indo-Aryan group[6] like

---

[6]It is said that this migration happened in the 12th century BC. The atrocities of the Muslim rulers in India were the main reason for migration of these

Brahmins and Chettris from the present-day India also affected the Begnas as these people with their advanced technological and social and political organization gradually took over not only the resources but also the political power. The chieftainship of Begnas as in other micro-states was taken over by the Chettris. The existence of a fort at Begnas subsettlement, which was located at the highest altitude of the village, and the presence of occupational castes and Ghartis shows that the former kingdom was ruled from this spot. It was said that members of the occupational caste and Ghartis (who were slaves before their freedom in 1926) were brought by Brahmins and Chettris in Nepal for various services. Even though most place names, river names and field names were derived from Magar language, the population of Magars in Begnas was small. The main reason for this as told by elderly citizens of the village was that a plague had hit in places, mainly the Lamaswara, where Magars were concentrated. This plague also caused a heavy loss of lives. The surviving Magars had also migrated to other places because of the fear of the plague. Later on, Brahmins, Chettris, Thakalis and Gurungs occupied the Magars' land.

The above discussed historical process brought people of diverse origin in Begnas. Even though it appeared that the residents form a homogeneous group, the caste system was deeply rooted and it had hierarchically stratified the society.

---

people who were the followers of the Hindu religion. At first, they came to the western Nepal and from there they slowly migrated to the eastern regions. It is also said that these Indo-Aryan immigrants introduced the caste system.

The majority followed Hindu religion and only about eight per cent were Bhuddists. There were three houses of Muslims who came here to do business. These Muslims were concentrated in the market area. Muslim lied outside the hierarchy of the village society and thus they were not bounded by social obligations. Followers of the Buddhist religion were mainly the Magars and the Gurungs, but they were also placed in the caste structure even though their original religion was shamanism akin to Bon religion of Tibet[7]. Brahmins and Chettris comprised about two-thirds of the total population, while Gurungs and Magars made up 16 per cent. Occupational castes, which included *kami* or blacksmith, *damai* or tailors, *pode* or fishermen and *sarki* or cobbler, made up about eleven per cent of the population. These castes were considered to be untouchables. Brahmins and Chettris were concentrated in Begnas and Lamaswara, and Gurungs and Magars in Ward 8 and at Bhangara. Occupational castes were distributed across the subsettlements, but *pode* were settled near the lakes. Thakalis and Newars were the recent immigrants in the village. They were involved in a small business in the area and most of them were settled at Pachabaiya. Lamas, who also practiced, apart from farming, exorcism, were also present in the village.

---

[7]In the anthropological literature of Nepal, caste and ethnic groups are generally used synonymously. In recent times, various minority groups have shown concerns in the use of caste and ethnic groups interchangeably. They want to use ethnic groups for the minority groups, particularly those outside the Hindu religion.

Due to the existence of the caste system and people not belonging to Hindu religion, multiple (social, ritual, economic and political) relationships had developed among caste and ethnic groups. People understood their social identity and status in terms of their caste status. Higher the caste of a villager, the more concerned he/she was about the caste status. It was due to this reason that villagers would at first like to know the caste status of an outsider. The occupation, marriages, food regulation and commensality were also guided by caste rules. Even though relaxation as to the occupational pattern could be seen, caste rules for marriage and commensality were still observed strictly.

According to the caste hierarchy, Brahmins, the priests, occupy the highest strata followed by the Chettris, the rulers and administrators. Then come Vaisyas, the businesspeople and farmers. Sudra, people providing service, occupied the lowest position. The untouchables remained outside the hierarchy. There was still controversy as to which castes belonged to the Vaisya and Sudra. Brahmins and Chettris wore the sacred thread. After the thread wearing ceremony, they would not eat food cooked or touched by members of lower castes. They also would not take meat, except for mutton, eggs and alcoholic drinks. Caste rules also forbade Brahmins to keep certain animals like pigs and chickens and to perform certain farm work like ploughing using animals. Members of the lower castes, i.e., occupational castes, were forbidden to enter the houses of the members of higher castes. Most of these castes should practice endogamy and hypergamous was not accepted. Despite these rules, some exceptional cases of marriage had been occurring in the study area as a result of which new castes with different roles and status were being created. This system of creating new castes helped to accommodate the persons and their children involved in such exceptional cases in the caste hierarchy. Other criteria of classifying people in the study area were `acceptability of water' and `drinking of alcohol', but these criteria were also implicitly linked with the caste rules.

Even though the caste distinctions were abolished in Nepal according to the Muluki Ain (civil code), 1963, the caste rules were still observed in the study area. Even the members of the lower caste, though aware of this legal provision, were also careful to observe the hierarchical separation imposed on them. But despite the observance of caste rules, members of one caste ridiculed the other caste and it also often led to inter-caste conflicts. Sometimes, tensions would also grow if a member of the lower castes threatened the status of a member of higher caste. In such cases, group division was based distinctly on the caste line. Outcasting was also commonly done to a member who brought shame to the caste by not observing the caste rules. The family and social lives of the outcasted persons were difficult. In recent past, as reported by the village residents, outcasting had not been done even to people not following the caste regulations, mainly because of less attachment to religious norms.

The cultural context of the study area, mainly the caste regulations for the conduct of an individual determined the inter-caste relationships which were also important in farming and in the management of common resources. For the formation of work groups or associations or for the exchanges of resources, these relationships played a vital role.

Families were the basic units of the village society. The basic structural lines of the families were formed by husband and wife (or wives, in some cases), children, daughter(s)-in-law and their children and any other relatives residing with them permanently. Even though, all family members contributed according to their age, strength and ability, the roles and status of the family members were also different. For example, daughter(s)-in-law were generally given a large share of the family work and usually their voice was not heard. Their social status was also low in the family. But, they generally conveyed their voice through their husbands. This sometimes created friction in the family, which eventually led to separation of the married children from the main family. Similarly, daughters did not have greater say in family matters, particularly in financial matters, as they were considered as the ones to go to other's home. The position of daughters further went down once they got married and became daughters-in-law. Family responsibility was transferred to the first born son or to other sons but not to daughters. The parental property was divided equally between all sons. A daughter would get a share on the parental property only if she remained unmarried and became more than 35 years age. Daughters were also least consulted in the acquisition or the disposal of the property. For the farm activities, advice of all the family members was heard and the decision was generally taken by asking all family members. The social place of older people in family was high and their advice was generally sought before taking decisions. The type of family in the study area ranged from a single member family consisting of a widow or a deserted wife to families having members of three generations.

Beyond the family, there was the lineage group called *kul* which included all families whose men could trace a patrilineal relationship to a common ancestor. Periodically they gathered in a place, *kul's* sacred place, to worship their lineage deity. This bond was not usually called into action. But only in extreme cases, which required co-operation or mutual help, this relationship was utilized.

The relationships between villagers, often across caste lines, were also developed to solve the common problems. For example, conservation of forest, pasture, school development, road development and like had given opportunities for villagers to work in groups. Similarly, many types of associations - often transient, formed and ceased continuously - between two or more families were formed depending upon their problems and mutual interests. These issues and their effects in decision-making of different groups of people are examined in Chapter 4.

C. The socio-economic setting

Population structure:

A large proportion of population of Begnas consisted of people below 15 years of age. The male and female population was almost equal in all age categories. Household size ranged between one and 18 members with an average of 6.53 persons. Farm size groups and ethnicity were directly related to the household size. The average household size of landless farmers was found estimated at 5.03, whereas that of large farmers at 8.73 individuals; it increased as the farm size increased. The reason for this positive relationship between farm size and family size was obviously the labor requirement for farming. As households with large farm size required more labor, they showed the tendency to have large family size because the family labor was considered to be the most reliable one. Due to availability of job opportunities in non-farm sector and the decline in the traditional forms of landlord-labor relationships[8], it was said in villages that only those families with enough family labor could only carry out the farming. As members of the higher castes had large land holdings in comparison to other households, they maintained a large family size.

Table 2.1: Respondents' average family size according to farm-size categories in Begnas in 1986/87.

Farm-size categories	Number of people in a family
Landless (n = 39)	5.03
Small (n = 79)	5.54
Medium (n = 92)	6.70
Large (n = 55)	8.73

(Source : Household survey, 1987)

[8] 1. The slavery system under which Ghartis used to work in the households of higher castes. This system was abolished in 1926.
2. Employing members of the occupational castes on a yearly or a seasonal basis.

The average number of children below 15 years of age in a family was estimated at 2.4 and that of old dependents above 60 years of age at 0.47. On average, dependents over 60 years of age in a family accounted for slightly more than seven per cent of the average household size. The dependency load[9] stood at about 45 per cent for the average household size. Therefore, the size of labor force comprising of the adult members of 15 to 60 years of age in a family was estimated at 3.57 individuals, i.e., 54.7 per cent of the average household size. Labor force size was largest among larger farm sizes, but the ratio of labor force size to family size was larger among the landless families.

Literacy and education:

About two-thirds of the respondents were found to be literate. The average educational attainment of respondents was slightly above primary schooling level, as the average educational score (AES) was estimated at 1.2 (1.0 AES means primary schooling). Only less than ten per cent of the respondents were higher educated. By caste, Brahmins and Chettris were better educated than members of other castes. Respondents' education showed a direct relationship with farm size categories. It was also observed that educational status of a family having an educated family head was also considerably higher. The majority of the respondents (82%)

---

[9]The dependency load in a family indicated the proportion of family members below 15 years and above 60 years of age. But in Nepal, particularly in the field study area, as observed in 1987 and 1993, these members also contributed to the family and farm works.

had sent their school age children to school. Proportionately, a large number of families having high educational attainment and owning larger size farms had sent their children to school. The reason as reported by the majority of the respondents was to enable their children to get other kinds of job, particularly the white-collar job. A very low percentage of the respondents were of the opinion that education is important for people engaged in farming. Even though quality of human capital rather than land is now considered important in farming, educated members in Nepali society have not been bringing positive changes in agriculture.

Since children were economic resources, the households not only lost the opportunity cost of children by sending them to school, but also had to incur additional investment. To cope with this additional burden, farmers had also brought changes to their farming. Major changes observed were disposing of some part of the land through mortgage or sale and hiring more laborers. Only a few of them grew cash crops to raise money for this purpose. The system of sending children to schools had aggravated the problem of labor shortage in village Nepal. Many families had to hire labor to compensate the labor shortage caused by schooling of the children. A large number of families had to look for cash income from non-farm work. As non-farm work was not available within the village, this necessity compelled many people to move away from home on a seasonal or temporary basis. Even though schooling up to fifth grade is free, it was a hardship for low-income families, especially of occupational castes, to send their children for higher education. Only families with large land holding or external income could afford to do that which eventually enabled their children to take up most of the leadership positions and to exploit new income opportunities.

The economy of the village:

The village economy was still based on land resources and it employed more than 80 per cent of the population of the village. Apart from farming, people also had some non-farm occupations. One third of the families had some of their members engaged in one or several types of non-farm occupations within or outside the country. Most of the Gurung and Magar families had their members engaged in the Military service in India. Some Chettris were also found to enlist in the military service in Nepal. Members of the occupational castes were engaged in professional work like tailoring, carpentry, blacksmithery and masonry. It was estimated that 20 per cent of the population was engaged in off-farm work, which required migration of people away from the village. About three fourth of the migrating workers worked outside the country. Members of the lower castes were also found to seek seasonal jobs in Pokhara. Off-farm income was vital for villagers to buy consumables not produced locally. Even food was purchased from the market. A large proportion of off-farm income was used for this purpose. A small part of earnings from outside was also used for redeeming the debt. This problem of redeeming loan was found to have compelled a few people to migrate. A few families having a large amount of savings from outside work also invested it in land and houses.

The farming in the village, which was subsistence-oriented, was operated on an average farm of less than a hectare. The household survey conducted in 1987 revealed that a household, on average, cultivated 19.8 *ropani* (1.00 ha) land. But a study conducted in 1992 (Pandey, 1992) revealed that average household owned 0.83 ha of cultivated land. Larger farmers, on average, had land holdings five times larger than the land holdings of small farmers. The difference was bigger in lowland holdings than in upland holdings. The average size of upland holdings was 9.94 *ropani* and that of lowland holdings was 9.92 *ropani*. Upland holdings were largest at Begnas, followed by Bhangara, because these settlements were located at high altitude and, thus, had proportionately vast upland areas. Pachabaiya being located at the lowest altitude had smallest upland holdings. Inversely, it had the largest lowland holdings. This shows that the location of the settlement was also related to types of land holding.

The land distribution in Begnas VDC was found to be inequitable. A majority of the families belonged to small and medium farm size groups. Landless and small farmers, who comprised about 45 per cent of the respondents, owned only 16 per cent of the village land, whereas large farmers, who accounted for only 21 per cent of the respondents, possessed 47 per cent of the land. Brahmins and Chettris were able to control proportionately more land than members of other caste farmers.

Families in Begnas had faced the problem of insufficient food production. In 1987, more than half the respondents reported that their food production was not sufficient for family consumption. For all the landless, for three fourths of the small farmers and about half of the medium farmers, food production was not sufficient. Comparatively, Brahmins and Chettris were sufficient food producers rather than the members of other castes. The data kept in the then Panchayat Office showed that food production was not sufficient for 40 per cent of the families for six months.

The main cause of food insufficiency lied in the small size of land holdings which was further diminishing because of increase in population and inheritance of the land by all sons and unmarried daughters. About 30 per cent of the holdings were too small to provide enough food. Another 40 per cent of the land holdings could produce sufficient food in a normal year only. Thus, the majority of the farmers had very little or no margin over the subsistence need. Hence, they may be hesitant to take risks in any kind of decision that may threaten their food sufficiency, however, little risky that might be.

Rice or paddy was the main food item that the respondents were found purchasing to meet the family food requirement. About one-third of the families had purchased rice or paddy. The other main items purchased were maize and millet. All landless and a majority of small farmers purchased rice or paddy and maize. Relatively very few farmers had sold their produce. About one-fifth of the households sold paddy, rice and millet. Very few farmers sold vegetables and fruits. This shows that in a community of largely subsistence farmers there were only a few surplus producer farmers who sell their produce.

Table 2.2: Distribution and size of land holdings in Begnas in 1987 (Land holding in *ropani*).

Farm-size categories	%age population	%age Land Occupied	Av. Upland Holding	Av. Lowland Holding
Landless (< 5 *ropani*)	14.7	1.5	1.3	1.7
Small (5 < 15 *ropani*)	29.7	14.5	4.7	4.8
Medium (15 < 30 *ropani*)	35.0	37.3	10.5	10.1
Large (30 or more *ropani*)	20.7	46.7	21.6	21.9

(Source: Household Survey, 1987).

In this study area, where a major portion of the food produced was used for family consumption, the decisions on which crops to grow might have been influenced by farmers' deliberations on which combinations of crops are most likely to provide sufficient food for their families. Surplus producers might have taken their personal preference as the basis for choosing the crop mix. Most of the surplus producers belonged to large and medium land-size categories and had comparatively high educational attainment. They were mostly Brahmin or Chettri castes.

Despite the existence of small land holdings, land leasing was also common in the study area. Three types of land leasing, viz: renting (on a fixed rental basis), share-cropping and cultivating *guthi* land (land in the name of religious institutions) were found to exist in the study area. Renting was not commonly practiced as only two respondents cultivated land on a fixed rental basis. In fact, they had rented land from a school. *Guthi* land cultivation also was not widely practiced. By and large, share-cropping was the common practice. The average size of rented land of all types was 1.48 *ropani*, i.e., about eight per cent of the average land holding size. Only families with large farm size rented out their land. Small and medium farmers were the ones to rent (in) the land, mostly on a share-cropping basis. Landless households generally did not get land for renting or share cropping because they lacked resource like animals and manure. Because of the availability of non-farm work, they were also not inclined to rent land.

In the study area mainly owner-cultivators had carried out the farming. In 1987, more than three-fourths of the families were owner-cultivators. There were only a few tenant cultivators in the village. About one-fifth of the families were owner-cum-tenant cultivators, i.e., they also cultivated other people's land. Owing to the large proportion of owner cultivators, the analysis by tenurial status did not show any significant differences. Land generally rented out was lowland located at some distance from the settlement. Upland land was rented out less frequently as it was not generally preferred because of low return from the cultivation of such land. More than four-fifths of the farmers, who had rented land, preferred to cultivate land on a share-cropping basis. The preference for a fixed rental basis was very low. The reason, as they said, was the 'high chance of damage of paddy by hailstorm'. Due to such uncertainty, they preferred to share the produce, generally at the ratio of 50 : 50.

Villagers purchased their daily necessities in the Pachabaiya market center. Basic consumer goods like rice, clothes, salt, medicines and soaps were available in this market. There were one or two teashops in each ward. Within the village boundaries, the non-market exchange system was still existing, e.g., exchanging labor for food or one food item for another food item. Exchanges of seed of one crop or variety of a crop for seed of another crop or another variety of the crop were very common. This eliminated the burden of producing and keeping little amounts of seed of all crops and varieties required for a household, because production and storage of seed needed special care and management.

Consumer goods available in the market of the village were supplied from Pokhara Town. There were less things that villagers offer to sell in the market. Few surplus producers offered rice for sale in the market. They sold maize, millet and most of the rice within the village through informal arrangements. Some people of Pachabaiya and Lamaswara offered vegetables like cowpea, cabbage, cauliflower and summer vegetables as well as fruits for sale. *Ghee* (produced after churning the butter; *ghee* has no water content) was the main commodity sold by villagers.

Institutional facilities in Begnas included schools, health posts and other governmental organizations to provide services like credit, technical know-how for farming including fish culture, electricity and irrigation. There were four primary schools and one high school in Begnas in 1987. Small Farmer's Development Project (SFDP) provided credit on a group- guarantee basis for small and landless farmers. The Fishery Development Office (FDO) provided technical know-how for fish farming. Two health posts located in Begnas provided traditional Ayurvedic (or herbal) and as well as modern allopathic health facilities. Electricity had been supplied in Begnas and a large proportion of houses were now connected with the electricity. One post office located at Pachabaiya provided formal communication facilities to villagers. Milk collection center established by the Dairy Development Corporation in 1990 had boosted the marketing opportunities for milk. Coffee was introduced in the village by the Government Project (Begnas Tal Rupa Tal Project - BTRT Project), but due to marketing problems, it was not popular among villagers. Since the early 1990s, villagers also formed producers' associations to carry out conservation, social and marketing functions. But these associations had not been successful in achieving their objectives. These were the modern institutions developed with the external agent, i.e., BTRT Project.

CHAPTER THREE

RESOURCES AND FARMERS' DECISION-MAKING

The main resources at the farmers' disposal were land, livestock and forest. The farming system in the study area had evolved or institutionalized as a result of the interaction between man and these resources. The role of ecological factors or the physical environment in shaping this farming system is discussed in this Chapter. How the given ecological environment (local resources) had been managed (until the mid-1990s) and utilized by farmers vis-a-vis their objectives is the especial concern of this Chapter. Since a significant part of the village area was also covered by water, the use of this resource has also been discussed. This Chapter is, thus, divided into four main sections: management of agricultural land (cropping pattern), livestock, forest and water resources.

A. Cropping pattern

Farmers of Begnas cultivated a variety crops in order to utilize the diverse nature of the environment and to fully utilize the family resources with the aim of meeting the food requirements from their own farming. Even though a large number of families in the study were then not able to fulfill their food requirements from their farm - the discussions in the preceding Chapter also suggest this - the existing farming system had been oriented in that direction. Paddy, maize and millet were the main crops grown in the village. Paddy (wet) was grown in lowland and the dry crops like maize and millet in the upland. About 90 per cent of the cultivated area was devoted to these crops. About one half of the total cropped area was utilized for paddy cultivation. Except for few absolutely landless households, which constituted about 2 per cent of the total households, all other households grew paddy, maize and millet. The total area devoted by an average household to wheat, pulse and mustard cultivation was minimal (about four per cent only). In 1987 almost the entire lowland area was utilized for growing only a crop of paddy. Other crops were grown in upland fields, except for pulse, which was also grown on the levees of paddy terraces. Mixed cropping was common in upland fields. The entire upland paddy was grown mixed with maize. Millet was inter-cropped with maize or transplanted as sequential crop. Cowpea and soybean were grown mixing with maize (Fig. 3.1).

An attempt was made to examine how effectively and efficiently upland and lowland were used by various types of farmers. To make the comparison easier, indices were developed to measure the intensity of cropping and the time period the land remained covered by crops. The former was measured by Multi Cropping Index (MCI) and the latter by

Land Utilization Index (LUI)[10] MCI indicates number of crops grown in the same piece of land in a year and LUI indicates proportion of time in a year over which crops occupy the land. MCI for the whole village was estimated at 1.32, which means that 32 per cent area was double cropped and the rest was single cropped. The average LUI was found to be 0.47, which implies that land was occupied by crops for 171 days of the year, i.e., 47 per cent of a year. MCI and LUI for the upland were estimated at 1.56 and 0.51 respectively. The same figures for lowland were estimated at 1.0 and 0 41 respectively (Table 3.1). This clearly indicates that the

---

10

$$MCI = \frac{\sum_{i=1}^{n} a_i}{A}$$

Where,
a = areas planted to i^{th} crops and harvested during a year.
A = total cultivated land area available for use during that year.
n = total number of crops.
1.0 MCI means that land is occupied by a single crop in that year.

$$LUI = \frac{\sum_{i=1}^{n} a_i t_i}{A.T}$$

Where,
a = total number of crops grown by farmers in time t.
T = time period under study (365 days)
A = total cultivated land area available for use during T.
n = total number of crops
LUI represents the portion of a year time over which crops occupy the land.
1.0 LUI means that all the area is covered by crops throughout the year.

Fig. 3.1: Predominant cropping pattern in the study area (Begnas) In 1986/87.

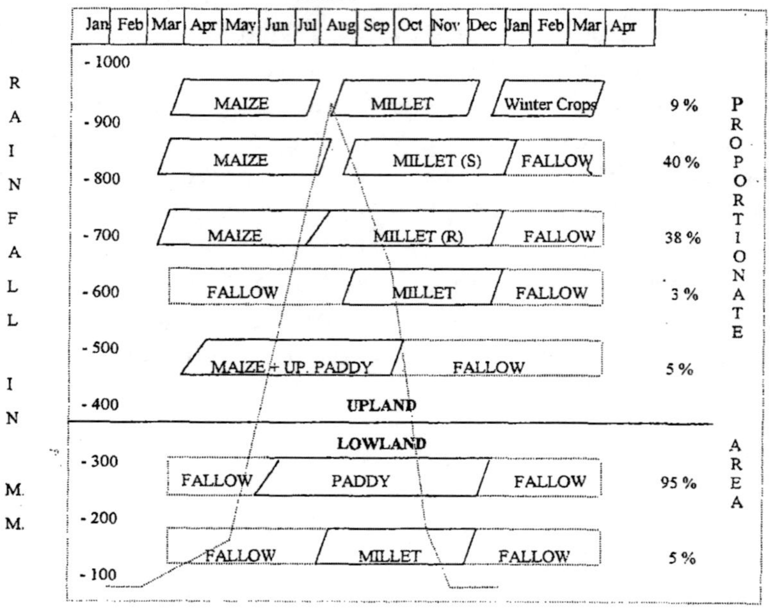

NB: The percentage on the right hand side shows the area occupied by the corresponding cropping pattern. These figures do not add up to 100 % for each land type, because of the exclusion of minor cropping patterns.

Rainfall is the average amount of five years (1980-84).

UP. PADDY = UPLAND PADDY, MILLET (S) = MILLET SEQUENTIAL,
MILLET (R) = MILLET RELAYED

upland area was better utilized than the lowland area. The reason for better utilization of land in upland areas was related to the location of the settlements (see the preceding Chapter). Since settlements were located in upland areas, manure was produced and utilized in fields of these areas. The collected animal manure was difficult to transport to the lowland, due to distance and rough geophysical conditions. Therefore, the villagers followed a different technique of manuring the lowland, i.e., by allowing the animals to graze there in the winter season, which also put limits to better utilization. But in upland areas, individual plots were fenced by stonewalls or bamboo to protect crops from foraging animals, whereas in lowland areas such individual fencing was not done. Thus, society had control to some extent in paddy cultivation, whereas more individualization was evident in upland farming. Individualization permitted small and landless farmers to put extra efforts in crop cultivation in their upland fields. In their lowland fields, they were somewhat inhibited to grow more crops. High MCI and LUI in case of small farmers in comparison to those of medium and large farmers, was due to their efficient crop production in their upland fields. But by 1993, after six years of previous field research, rapid changes in cropping pattern were brought about by villagers. Intensity of cropping was rapidly growing in lowland, a large part of which was used to grow three crops.

Table 3.1: Cropping pattern in Begnas : area devoted to different crops by an average family (1986/87).

Crops	Av. area (*ropani*)	Crops	Av. area (*ropani*)
Lowland paddy	12.22 (47.9)	Upland paddy	2.07 (8.1)
Maize	4.83 (18.9)	Millet	5.14 (10.1)
Pulse	0.50 (2.0)	Wheat	0.33 (1.29)
Mustard	0.43 (1.7)	-	-
Total cropped area	25.52 (100.0)	Av. land holding	19.36(100.0)
Total cropped upland	15.53 (60.5)	Av. upland holding	9.94 (51.3)
Total cropped lowland	9.92 (39.2)	Av. lowland holding	9.92 (48.7)
MCI total land	1.32	LUI total land	0.47
MCI upland	1.56	LUI upland	0.51
MCI lowland	1.00	LUI lowland	0.41

(Source : Household survey, 1986/87; Figures in brackets are percentages; Av. - average)

Farmers' practice of using collected manure only in upland fields was quite appropriate in order to increase the water retention capacity of the soil, which permitted farmers to grow crops without irrigation. Otherwise, farmers would not have been able to grow crops in upland areas. Use of more farm manure in upland fields also seemed sensible from the point of view of replenishing the nutrients lost by soil erosion. As the terraces in upland were not well maintained and were slightly outward sloping, rate of soil erosion was high in such fields. But it was also necessary to keep terraces in this condition in order to avoid water-logging which was detrimental to the growth of dry crops like maize and millet grown in upland. Moreover, soil erosion was naturally higher in upland than lowland because of higher slopes. But the traditional practices of growing various types of fodder trees and grasses in the levees of upland terraces helped to reduce the erosion rate in such land. In steep areas farmers also kept large levees covered by grass which acted as buffer zone between terraces.

Table 3.2: Percentage of farmers (farm size categories) growing different types of crops in Begnas (1986/87).

Farmers	Lowland paddy	Upland paddy	Maize	Millet
Landless (n = 39)	37.2	50.5	76.6	77.4
Small (n = 79)	92.4	65.8	88.6	84.8
Medium (n = 93)	98.9	36.6	95.7	94.6
Large (n = 55)	100.0	70.9	98.2	94.5

(Source: Household Survey, 1987; Multiple Choice Question).

The cropping pattern varied among different land holding categories (Table 3.2). Proportionately, more large farmers grew paddy, whereas small and medium farmers grew millet and maize. At the same time, large farmers devoted proportionately more land to paddy production, whereas landless and medium farmers to millet and maize production. The reason was that paddy cultivation was done in lowland fields, which were largely owned by big farmers, as there was more variation in sizes of lowland fields among different categories of owner farmers. Small farmers giving more preference to millet cultivation would also do so out of their dietary preference. People in the study area had the notion that millet gives more energy, and rice consumption is prestigious. This might have been a reason for small farmers, particularly the members of the occupational castes purchasing more millet despite its high cost. Moreover, millet was also used by members of lower castes households and Tibeto-Burman groups including Gurungs and Magars to brew alcohol. The caste regulations did not restrict them (also classified as *matwali*, alcohol drinkers) to produce, consume and sale alcohol. Due to some relaxation in caste rules on the consumption of alcohol, its production had also become an income source for a few families of the study area. The production of millet by large farmers, especially the Brahmins and Chettris, was intended to exchange this for labor and for feeding animals.

Variation in cropping pattern was also found to exist among the subsettlements. This variation was primarily due to altitude. More area at Pachabaiya was devoted to paddy cultivation because the average lowland holding was largest due to its low altitude, whereas millet occupied relatively large proportions of the areas at Bhangara and Begnas. New crops like wheat, vegetables and fruits were grown mostly at Pachabaiya. This may be attributed to its location near the market center, i.e., Pachabaiya-Khudimuhan. It was also observed that uneducated and poor households in every ethnic group were more concerned about their farming in comparison to village residents with higher education. The main reason for this was the prevailing notion in the community that educated persons should not perform manual work like farming. But some of the educated people were found to introduce new crops like coffee and other fruit and fodder trees including trees for aesthetic purpose. The productivity of all crops was found to be highest among small farmers. It was found decreasing as the farm size increased and, as demonstrated in the preceding Chapter, educated families had large farm size in general. But despite higher yields of crops, average production of small and landless farmers was largely insufficient.

Despite the finding of a positive relationship between education and productivity in the context of Asia by most researchers (Lockheed et al., 1987), this relationship was found not to exist in the study area. The reason may be related to Schultz's (1975) hypothesis that effects of education would be greater under modernizing conditions, which, in fact, attract educated people to do agri-business. Most of the research done in the African context also shows a negative relationship, between education and productivity.

Cash crops were not commonly grown in Begnas. Until 1993, only about one-sixth of the village households produced some cash crops for the market. By settlement, more than one-one-third of the cash crop growers were residents of Pachabaiya and the remaining of Lamaswara. By farm size, the highest proportion were small farmers, followed by those medium farmers who were producers of cash crops. Vegetables and fruits were the main cash crops grown by villagers.

The implication of this analysis may be that there was diversification in crop cultivation. Ecological factors helped to diversify the cropping pattern. Different crops grown in the farm were mainly intended to meet the total family requirements. Dietary preferences and location of households in relation to the market center were also found affecting decisions on crop choice. Farmers without probing deeply into the evolution of their diversified farming in the light of ecological diversity followed this system as tradition, which was locally called *chalan* (the way of doing things). *Hamro chalan* (our way of doing things) was the common answer villagers gave to the question `why they have adopted this type of farming practices?' Despite this, new crops and varieties had also been introduced in the village.

Terracing and soil conservation:

Farmers in the study area and in all hill villages of Nepal had adopted terraces for field agriculture, and there was a network of terraces even if the hill flank seemed to be almost too steep. These terraces were found not to be truly consistent with contour lines. It is also argued that lands with more than 27° slope should not be cultivated, but in the study area some fields had as much as 40-45° slope. The system of terrace cultivation had been minimizing the soil erosion that would otherwise have occurred to a much greater extent. It has been revealed from various historical records that the system of terracing was started from the 14th century BC when Hindu people belonging to Indo-Aryan race came here to settle from India. These immigrants occupied most of the valley floors and gently sloping areas where rice could be grown as they were accustomed with the use of hoe and the cultivation of rice. To grow rice in hilly areas terracing was vital for the accumulation of water and this led to the process of terracing of hill slopes. The original or the first settlers (members of the Tibeto-Burman group) who were believed to come to Nepal from Tibet used to practice shifting cultivation but as their interaction with the new immigrants increased they also gradually adopted terracing and settled agriculture. Some remnants of shifting cultivation could still be found in villages of Nepal dominated by these people. This account of terracing tells us that it has been followed as tradition, which has been developed due to historical and ecological reasons.

Farmers had a distinct system of ploughing these terraces. They ploughed these terraces in an upward and inward direction. The reasons in the native view were that, firstly, no soil slides from the edge of the terrace and, secondly, it was considered auspicious. Ploughing was done by a pair of oxen. The local measurement unit of land was *hal*. One *hal* was equivalent to the area ploughed by a pair of bullocks in one day. Thus, the actual amount of area in one *hal* varied according to tillage method, soil and crop. Requirements of inputs like human and animal labor, seed rate and manure were calculated on the basis *hal* mainly because it took into account only the land which was cultivated, i.e. excluding the area occupied by bonds or levees and the uncultivated area in between terraces. Quality and types of land was also taken into account while calculating input rates per *hal*. For example, more seed rate was used in less fertile land. It was reported that about 0.3 kg seed of rice was sufficient for a *hal* of irrigated paddy land but this rate was doubled for the rainfed paddy land. Farmers before planting crops, scarped the levees to remove weeds grown on the surface of such levees and this process helped in minimizing the attack of insects and pests and in increasing the fertility of soil in terraces.

Farmers used very simple tools and equipment for cultivation mainly because of the small sizes of terraces. Iron and wooden plows were used for ploughing. *Kuto*, *kodalo* and *bauso* were used for digging purposes, and *ansi* and *khukhuri* were the tools for cutting purposes. People rarely used machines for the farming purpose. Small terraces and steep slopes also precluded the possibility of mechanized farming. It was one of the reasons that farmers were not adopting line planting of crops in Nepal because it was appropriate and cost efficient only if machines were used for weeding and harvesting.

Diversified cropping pattern and labor use:

Because of the multiple and diversified cropping pattern, a farm family grew different crops in a cycle. The cropping cycle followed by farmers of Begnas is shown in Fig. 3.2, which also shows the timing of different crops. It shows that farmers had to be quite busy during the months of

May through November. The relatively slack months for the farmers were from February through April (Figs. 3.3 and 3.4). However, other non-farm activities were carried out during this time. Females spent a significant portion of this time collecting firewood. Also, many of the household heads engaged in non-farm work in order to earn cash income to purchase supplementary food, farming items and other things. The farmers and their families were also engaged in house repair, looking after the cattle, carrying compost and doing other works during this period. In conclusion, the farmers and their families were quite busy throughout the year either in crop production or in other works. There was a critical labor shortage during paddy transplanting. Another season of peak labor demand was during harvesting of paddy. Uncertainty of hailstorm and rain forced the farmers to harvest paddy as soon as possible after the maturation of the grain.

The mixed cropping tradition in both upland and lowland areas had the effect of easing labor problems, since differences in respect to the timing of high labor requirements allowed farmers to make use of family labor throughout the year. In other words, diversified cropping pattern helped farmers to spread the requirement of labor throughout the year (Fig. 3.3). This is illustrated in Figs. 3.3 and 3.4, which show that crop diversification had eased the problem of fluctuating labor requirements. These figures also show that the problem of labor shortage would have been worst if farmers had resorted to specialized cropping.

For example, Fig 3.4b shows that farmers would have to utilize their labor mainly in June to August and in November to December if they had grown only paddy. Figs. 3.4c and 3.4d demonstrate similar cases of fluctuations in labor requirements if farmers were to grow only maize or millet. But by combining all these crops, this fluctuations in labor requirements had been minimized (Fig. 3.4a). The low coefficient of variation of the labor requirement in one year for the entire cropping system, in comparison to that of individual crops, shows that the present system was better in terms of labor requirements than specialized cropping. The most labor intensive crop was rice, followed by maize and millet. Therefore, exchange labor was more common in rice cultivation. Labor inputs for major activities of different crops are shown in (Appendix) Table A.1. If other works like post-harvest activities, other minor cropping activities and livestock rearing were considered, the total labor input into the existing farming system would have been much greater than that shown in Table A.1.

To cope with the labor shortages and to perform certain work requiring team efforts, farmers were found to use hired and exchange labor (Figs. 3.5 and 3.6). A special attempt was made to study the labor utilization in rice, the main crop, cultivation in the study area. The household survey conducted in 1987 showed that about 40 per cent of the rice cultivation work was performed by exchange labor and 33 per cent by own labor. The remainder was performed using hired labor. Proportionately more hired labor was used for land preparation as Brahmins and Chettris were not doing this job due to caste restrictions. Male members of the occupational castes and Gharti were generally employed for this purpose.

Fig. 3.2: Cropping cycle in the study area (1986/87)

- Months -

Crops	Jan	Feb	Mar	Apr	May	Jun	Jul	Aug	Sep	Oct	Nov	Dec
Wheat						*	***	***	***	***		
Potato				**	***	***	***	*				
Mustard				***	***	***	***					
Up. Paddy	***	***	***								***	***
Winter Vegetable								*	***	***	***	
Summer Vegetable		***	***	***	***							
Blackgram		***	***	***	***	***						
Cowpea	***	***	***								***	***
Millet (S)				***	***	***	**					
Millet (R)		*	***	***	***	*						
Maize	***	***	***								***	***
Dry rice	***	***	***	***	***						***	***
Wet rice	***	***	***	***	***	***					**	***

NB: Up. Paddy = Upland paddy, Millet (S) = Millet Sequential, Millet (R) = Millet Relayed Dry Rice (Paddy) = Paddy grown in unirrigated land. Wet Rice (Paddy) = Paddy grown in irrigated land.

Fig. 3.3: Various production activities of different crops in a yearly time framework (1986/87).

- Half Months -

1	2	3	4	5	6	7	8	9	10	11	12	13	14	15	16	17	18	19	20	21	22	23	24
T			W	H	E	A	T								S	S		W				H	
					S	S		T	P		W				H		T	M	I	L	L	E	T S
			S	S			T	P		W				H			T	M	I	L	L	E	T R
W		H					H				M	A	I	Z	E					P L		S	S
	S S			P L		T	P				W				H		T	D	R	Y	R	I	C E
S S		P	L		T	P			W					W		H		T	W	E	T	R I	C E

Note : SS - Seed Sowing, PL – Ploughing, TP - Transplanting
W - Weeding, H – Harvesting, T - Threshing
Millet R - Millet Relayed, Millet S - Millet Sequential
Half Months start from the beginning of May.

Fig. 3.4a: Total labor inputs for all crops in an average farm (1986/87)

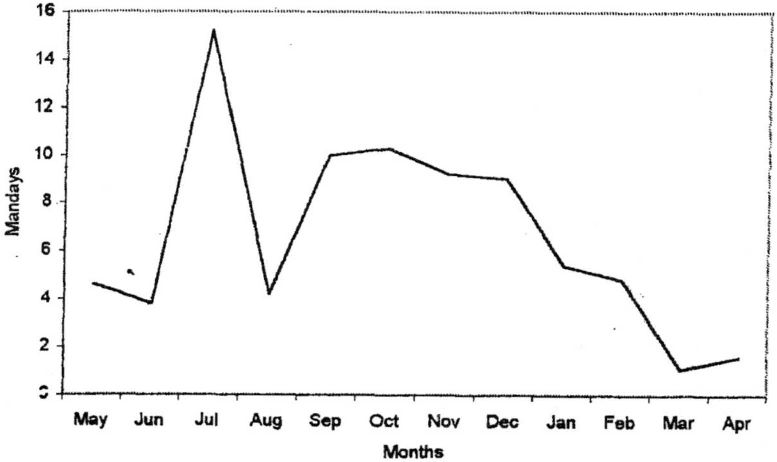

Fig. 3.4b: Labor input in rice cultivation in an average farm (1986/87)

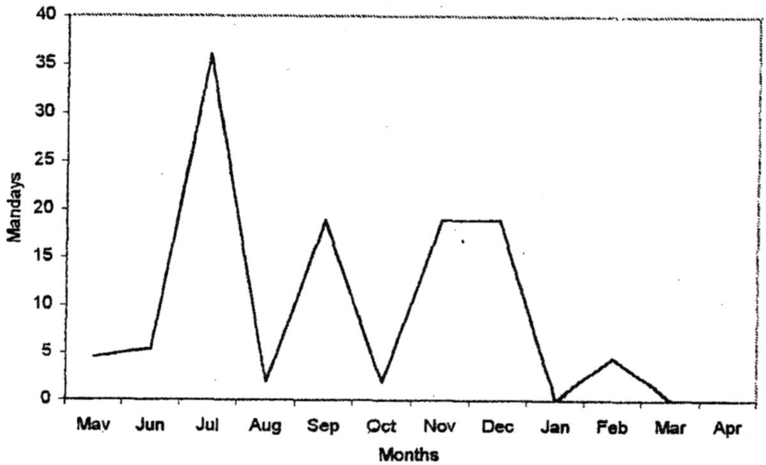

Fig. 3.4c: Labor input in maize cultivation in an average farm (1986/87)

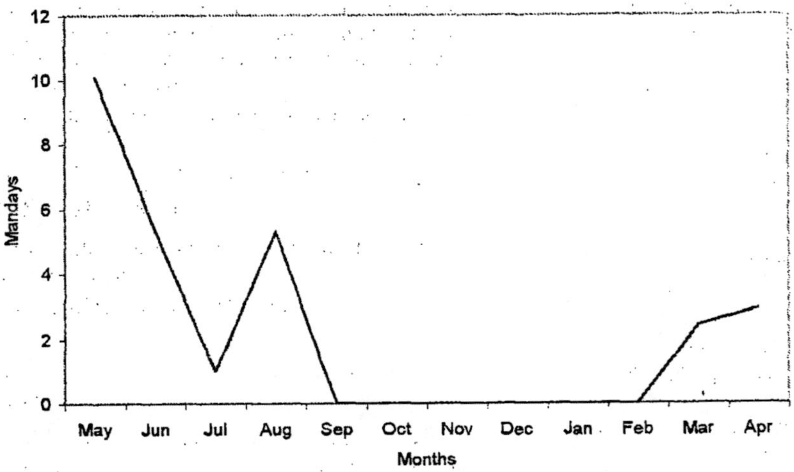

Fig. 3.4d: Labor input in millet cultivation in an average farm (1986/87)

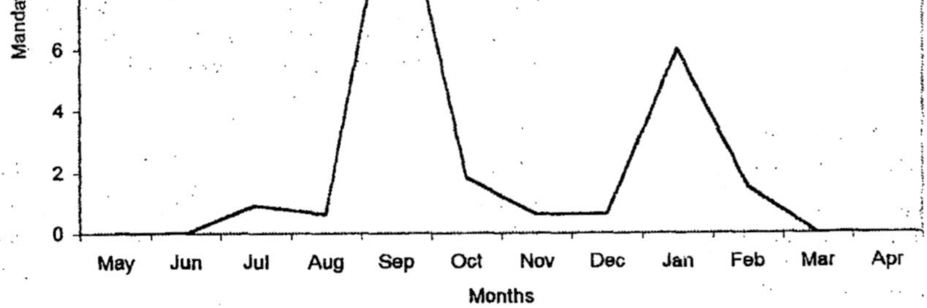

Fig. 3.5: Types of labor used in rice cultivation by farm size (1986/87)

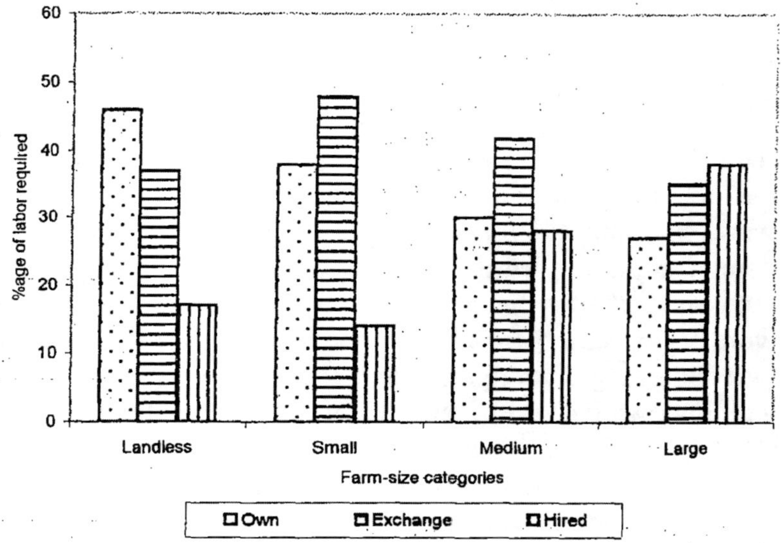

Fig. 3.6: Types of labor used in various activities of rice cultivation (1986/87)

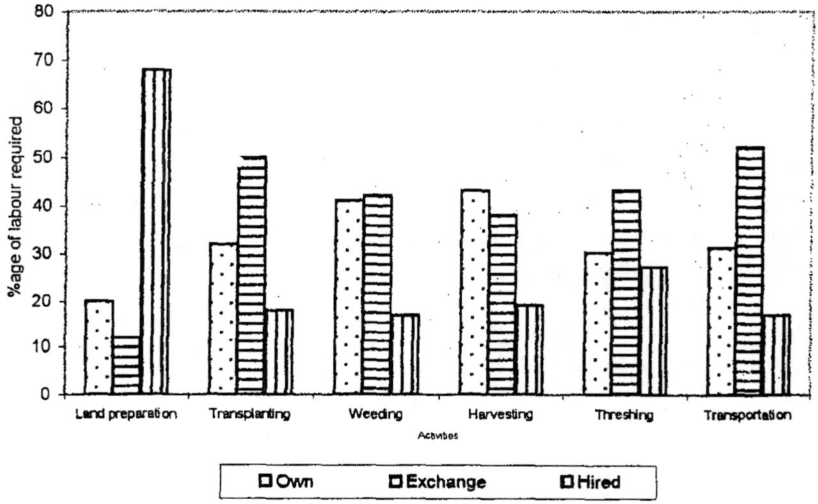

The type of labor utilized was also found to have depended on the nature of work. Most of the transplanting, threshing and transportation work was done using exchange labor. Since it was necessary to finish these works as soon as possible once they were started, exchange labor was utilized for these purpose. Since threshing of paddy was done in an open space, it had to be finished within one day, thus requiring much more labor at a time. The system of exchanging labor had been helping farmers to carry out such work smoothly. Works like weeding and harvesting were done using family labor. Since weeding could be done over a long span of time, generally family labor was used for this work. The reason for using more family labor in harvesting was stemming from the fact that every farmer would be busy with his own harvest because it should be completed within a short span of time. Therefore, it was difficult to get any labor for hire, unless laborers were contracted and paid in advance. During this time, all members of the family including children and old people worked from dawn to dusk. Large farmers were found using a high proportion of hired labor, whereas the practice of exchanging labor was found more common among small and medium farmers. Regarding the mode of payment to hired laborers, proportionately more respondents paid in both cash and kind. Purely cash payment was rare. Hired laborers were paid either by giving grains in advance, in most cases during festivals, or by giving their share of the labor output. The latter was employed in the case of harvesting operations. Still another method of hired labor employment was contracting labor for a year's work on a fixed agreement. Laborers were paid certain amount of grains [locally called *mujuri*; a laborer so hired would earn about 5-6 *muris* of paddy in a year from the family (families) using his service] depending upon their pre-fixed agreement. Generally, ploughmen were employed on such an annual basis.

Apart from spreading the labor requirement throughout the year, the mixed and diversified cropping strategy that farmers were adopting had other advantages, too. Differences in harvesting time provided food, while other crops were growing, thus alleviating the need for storing food for long duration, and also offered a variety of food. Further, these differences reduced the risk of hailstorm or pest damage. Apart from the main crops that were included in the cropping pattern analysis, every household maintained a sort of kitchen garden where few numbers of various crops were grown for home use. Generally these crops were gourd, pumpkin, tomato, pea, bean, chilly, turmeric, amaranth, hemp and various other seasonal crop.

The system of crop rotation:

Farmers followed the system of crop rotation which extended over one, two, or more years. They had also the practice of keeping their land fallow (see Figs. 3.1 and 3.2). Rotation of crops differed according to types of land. In *abal* (excellent quality) land, intensive crop rotation was followed without leaving the land fallow. The *char* (poor quality) land was often left fallow and then used as pasture. After two or three years of grazing, these lands were utilized for growing crops again.

Crop rotations were not kept the same year after year. Villagers changed the rotation, considering the changes in the yield of the crop. Farmers were aware that yield decreases if the same crop is planted every year. This should have been the case in such places where the originality of seedstock was not maintained, and yield decreased after some time due to complex genetic reasons. Moreover, soil would have become severely deficient in a particular nutrient if the same crop were grown again and again. In the light of such biological causes of decreasing yield and of soil nutrients, another tradition, which the farmers were following, was that they changed between seeds from other places. The variety, which deviated from normal characteristics, was locally called *thimaha*. If more such crops were seen in the field, they either changed the crop or changed the variety. A wide range of local varieties was found to have maintained within the village. Eighteen local varieties of rice, three of maize and eight of millet were recorded in the village (Appendix Table A.2). While selecting varieties, most of the farmers were found to give preference to the suitability of the varieties to the land cultivated by farmers. The aim of this selection criterion was to obtain highest possible yields. As a farm family owned different types of land, different varieties of crops were needed to be grown according to the suitability in different pieces of land. But farmers with large land holding were also found to give preference to palatability while choosing rice varieties. They did so because they were more concerned with the quality of rice as they did not face food shortages. Moreover they also had good land (with irrigation) to grow palatable rice varieties.

Risk and uncertainty and farming:

Many types of risks and uncertainties also affected farmers in Begnas. The farmers were more concerned with the risks arising from the climatic factors like rainfall, wind and hailstorm. Due to lack of technological means to deal with these uncertainties they had developed their own traditional mechanism to cope with such uncertainties. This coping mechanism differed from one crop to another depending upon the nature and agronomic practices required for the crop and the types of potential risks. For example, rainfall uncertainties were most critical for rice transplanting (discussed in Chapter 2). In some years, farmers were not able to transplant paddy in rainfed areas. As an alternative, they cultivated millet in some parts of such areas after the season for paddy had elapsed.

Hailstorm caused substantial loss in this area, causing damages both to paddy and maize crops and to fruit production. The critical period of damage to paddy was during November to December, i.e., the maturation time of paddy. Another critical period was June. Hailstorms caused maximum damage when the paddy was at the maturing stage and maize at the knee-height stage. To avoid this situation, farmers planted paddy about 15 to 20 days earlier, whereas maize was sown about a month late to make sure that its growth by June would not be more than half a foot. In the case of rice, some farmers preferred the *ramani* variety which was more resistant to hail damage but at the same time required more labor for threshing, because it was somewhat difficult to handle. Hailstorm was the main reason for the farmers' hesitancy in growing fruit trees despite favorable temperature and rainfall for the growth of subtropical and alpine fruits. Effects of hailstorm in villages of Nepal were also localized. Within the boundaries of a small village, some area remained not affected by hailstorm while others were severely affected.

Wind damage was critical to the maize crop during its kernelling period, i.e. when it started producing ears, which happened in August. Referring to this problem farmers said that they did not plant improved varieties of maize like 'Khumal' or 'Rampur Yellow', because these were more prone to wind damage owing to their tall growing nature. Wind made the plant lodge on the ground and such plants produced small ears. There was also the high chance of not having grains in the ears. Wind damage was also considerably high in the case of paddy and therefore very tall varieties of paddy were not preferred. At the same time, dwarf varieties of paddy were not preferred because they did not produce enough straw, which was used as animal feed. Millet was the crop least affected by such uncertainties. This might have been the reason for small farmers preferring to grow more millet.

It appears from the above discussion that strategies followed by farmers were all related to minimizing the risk situation. But it does not mean that they were totally risk averse. Despite the high chance of hailstorm damage besides other uncertainties causing losses, the key informant survey regarding the damages by hailstorm during the last ten years showed that there was a probability of 10.1 per cent damage to rice or maize every year. Despite this, farmers had been growing all kinds of crops without cessation (Appendix Table A.3) and in most cases they were striving to produce more. This was clearly illustrated from farmers' practice of planting of millet in some of the paddy area, when monsoon rainfalls were insufficient.

To cope with these uncertainties and risks, which were usually not in their control, ritual ceremonies were also conducted which differed from one crop to another and from one cultural group to another. This practice also affected the production activities and the social life of the villagers. It is covered in the next Chapter.

Changes in the cropping pattern:

Revisit to the study area in 1993, revealed many changes in the cropping system in Begnas. The major change was the decline in winter fallowing of lowland (***khet***) and introduction of not only new varieties of crops but also of new crops. The intensity of cropping in the lowland was found to have increased significantly[11]. This led to the decline in the *goth* system of livestock raising which has been covered in the following section. In most of the *khet* land in Begnas, three crops were grown in 1993, whereas just about six years ago only a single crop of paddy used to be grown. In lowland where irrigation was available in winter, early paddy (mid-March to mid-June) was grown which was followed by main season paddy crop (July to mid-November) and then by wheat (December to mid-March). Where irrigation was not available, main season paddy was followed by wheat and then by maize (mid-March to mid-June). Therefore, a large part of lowland was now covered by crops through out the year. In upland, different types of trees useful for fodder as well as timber were grown. Even members of the occupational caste had started growing vegetables for their own consumption, but this practice had caused them to abandon the practice of keeping pigs. Only a few households were found to keep pigs but now they were required not to let them forage on the field of neighbors. Crops like coffee, cardamom, pineapple were also introduced in the village but they were still in experimental stage and thus were not adopted at a full scale. Similarly fodder grass like 'Nepier' were also introduced and many farmers were

[11]Based on data collected from a household survey in 1987, Multi-Cropping Index (MCI) for lowland was estimated at 1.0. But the value of MCI as estimated by Pandey (1992) for the same area ranged from 1.45 to 1.18.

found to grow this grass in their barren land and in the levees of terraces. Cropping intensity in upland had also increased[12]. In addition to maize and millet, mustard was grown as a major crop. This increase in cropping intensity had been associated with the decline in the fertility of land. Villagers, by early 1993, had already felt this as a serious problem. They reported that use of chemical fertilizer had become a must for crop production. Moreover they were also finding the decline in the taste of food grains and vegetables for which they blamed fertilizer and other chemicals. Improved varieties of crops like maize, millet, rice and wheat were commonly grown nowadays. Due to popularization of few improved varieties of each crop, it was also likely that the rich genetic resources of the village, i.e., existence of many local varieties (see Appendix Table A.2), would be lost and farmers range of choice in this regard would be narrowed.

B. LIVESTOCK

Livestock raising was quite popular in the study area. No household surveyed was without some livestock. Livestock raising was well integrated with the cropping system.

---

[12] MCI in 1987 for the upland was 1.56. But the value of MCI as estimated by Pandey (1992) for the year 1991 ranged from 1.90 to 1.79.

Farmers' raised different types of livestock like cow, buffalo, bullock, goat, sheep, pig and poultry. The average livestock ownership was measured by Average Livestock Unit (ALU)[13] which helped to aggregate the number of different types of livestock owned by a family.

Average livestock ownership was high in the village, as the ALU was estimated at 5.5 heads (Table 3.3). By settlements, large numbers of livestock per family were raised at Bhangara (ALU = 7.8), followed by Lamaswara and Begnas. The reason for high livestock ownership at Bhangara was its closeness to the forest. The ownership of livestock was lowest at Pachabaiya, which was near to the market center. Land ownership was directly related to the numbers of heads of livestock raised. The number of animals (ALU) owned by large farmers was about four times larger than that owned by landless farmers. It was observed that Brahmins and Chettris owned more animals than members of other

---

[13] ALU is calculated on the basis of the following criteria:

$$ALU = \frac{1.5(B) + 1.0(C) + 0.4(S) + 0.2(P)}{N}$$

Where,

B = number of buffaloes,
C = number of cattle/ox,
S = number of sheep/goat,
P = number of poultry, and
N = B + C + S + P

Note: The small number of pigs was not taken into account.

castes. In some cases, wealthier families would also give their one or two animals to poor families, usually the members of the occupational caste, to be raised on a share basis. The tenant would get half of the incremental value of these animal(s).

The most commonly raised livestock species was buffalo, followed by goat. On average, one cow was raised per family. Ownership of livestock by species was found to be related to farm size categories and caste. Large animals, i.e., buffalo, cow and ox were mostly owned by large land holders, whereas there was not much variation in the ownership of small animals (goat, poultry and pig). The reasons for large farmers keeping large animals was that they needed them in farming as draft animals for power and for manure production; moreover, because of the large holdings, they produced more animal feed and fodder and thus could support these animals. Religious beliefs were reflected in the ownership of animals. Only members of the occupational caste owned pigs. Pens for pigs were built in their courtyards or were attached with the house. These animals being profane to practicing Hindus, members of higher caste did not keep them. On the other hand, cow being a sacred animal to Hindus, Brahmins and Chettris kept most cows. The purposes of raising large animals included home consumption of livestock produce, manure production and draft power. Consumption of dairy products was considered prestigious and as a result wealthier farmers tended to own more cows and buffaloes. Small animals were raised for subsidiary cash income required for the purchase of non-farm goods and services.

Table 3.3: Livestock ownership (ALU) in the Begnas VDC in 1986/87.

Settlement	ALU	Farm-size groups	ALU	Ethnic groups	ALU
Bhangara	7.8	Landless	2.3	Brahmin and Chettri	6.2
Lamaswara	7.0	Small	4.6	Others	4.8
Pachabaiya	4.8	Medium	6.0	-	-
Begnas	6.1	Large	8.4	-	-

(Source: Household Survey, 1987)

Ownership of oxen had a strong influence in social relationships and in production of crops as it provided power required in the farming system. Despite their use in ploughing, only about 43 per cent households owned a bullock. On the other hand only about 41 per cent households kept an ox (bullock). Farmers not owning any bullock were mostly small landholders, who exchanged labor for bullock power. Because of their small land holdings, they preferred not to keep any bullock, which otherwise would remain idle for most of the time. This exchange system also relived them of the burden of keeping one large animal. Farmers having an ox co-operated with a neighbor owning an ox so as to form a team. Therefore, their decision on the timing of ploughing was made through mutual agreement. This agreement for sharing oxen was renewed every year, but in some cases, the same team was maintained for a long time. This system maximizes the use of oxen. The problem of bullock shortage usually arose during the time of transplanting paddy. Farmers having no bullock had to delay their transplanting till others finished their work. Farmers with two or more oxen did not face this problem of mutual agreement and they could plough their land whenever they wanted. Only about 20 per cent of the households in the study area had two or more oxen and these farmers were considered landlords in the village.

Animals were kept in sheds, which were generally built very close to the farmhouses. Farmers made a temporary shed called *goth* in the paddy field after the harvesting of paddy, and they kept some of their animals in this place from November to April on the fallow used as pasture. This also was the common manuring practice in paddy fields. This spared the burden of carrying manure from the farmhouse to lowland fields. But, at the same time, it acted as a constraint on growing more crops. *Goth* had also some effects on social life. An adult male from a family went to stay in the *goth* every night. Kawakita (1977), studying a Magar village, found the tradition of sending young couples to stay in the *goth*, which he considered a tradition of minimizing vexation within the family. But in this study area, where Brahmins and Chettris formed the majority, only a male family member went to stay in the *goth*.

The main changes observed in the livestock sector in 1993, after six years of initial fieldwork, were the decline in the *goth* system and the introduction of exotic breeds of animals, particularly the cattle. Because of growing practice of cultivating winter crops in lowland, the number of households shifting *goth* there had diminished significantly because they had to look after their animals and graze them only in vacant land. The owners of animals were required to pay fine if their animals destroyed others' crops. Animal prisons were also built in the village and stray animals were taken to such prisons. Since plots in the *khet* (lowland) were still not fenced, more labor was required to look after the animals. Moreover, the practice of sending children to school had resulted in a serious problem of family labor shortage. It was also one of the reasons for keeping less animals. But the attention had shifted to keeping improved breed of animals. Improved breeds of cows like Holstein and Jersey could be seen in the village. To cope with the decline in the fertility of *khet* land because of the winter cropping, use of chemical fertilizer had increased. Some farmers with large family size kept a permanent *goth* in the *khet* and a few animals were kept here stall-fed throughout the year. Manure produced from these animals was used in the *khet*. These farmers also sold manure to others.

The rapid increase in prices of fertilizers since the early 1990s owing mainly to the government's liberalization policy discouraged farmers to use them. The prices of most of the fertilizers were increased by approx. three times. Even though, farmers were reluctant to use fertilizers, they were compelled to use them as the soil fertility had declined. They expressed that without fertilizer they could not grow crops now. But it was not so in the past as only local animal manure was sufficient and the soil quality preserved. Similarly, the use of pesticides and insecticides had increased significantly. Farmers were witnessing new types of diseases and pest in recent times, the control of which required new chemicals.

C. Forest

As stated earlier about 29.4 per cent of the village area was covered by forests. Even though trees grown in tropical as well as subtropical climate were available in the forest, *Schima wallichi* and *Castenopsis sp.* were predominant. According to the survey of BegnasTal RupaTal Project conducted in 1985, 23.1 per cent of the forest was classified as dense forest, 31.5 per cent as the open forest and 45.4 per cent as scrub.

In the past, the village forest was owned and managed on the basis of caste and lineage. For example, in Begnas subsettlement, lineage groups like Adhikari, Paudyal and Kandal owned forest. In Lamaswara, Tiwari, Kandal and Neupane controlled a large portion of forest in the name of their lineage and clan. They owned separate patches of forest, which they managed and utilized on the basis of norms developed by the consensus of the group members. Members of the lower castes did not have forest in their own name but were allowed to use the forest for fuelwood and fodder. The relationship they had developed with the members of the higher caste by providing caste service was beneficial for them in the sense that they were allowed to use the forest without any obligation of sharing the cost of management of the forest. The members of the higher castes also claimed that a part of the forest was registered in their names and they used to pay taxes for this ownership to the government. They continued to pay taxes till 1976 even though the forest was nationalized in 1965. They stopped paying taxes for the forest after the cadastral survey of land in 1976, which did not recognize the ownership of private forests. Since then, forest was used indiscriminately and a large proportion of accessible dense forest was converted to shrub land and grazing field.

Decline in the quality and quantity of forest was a hardship for villagers in general. But the members of the occupational and landless families were hit hard by this problem. Farmers with large land holding usually had many trees in their farmland and they could fulfil a part of the fuelwood and fodder requirement from their private land.

Forest in the Begnas VDC occupied land that could not be easily converted to crop land because of steep slope. But a few patches of land in Pachabaiya were clear felled for agricultural purpose and less sloppy land was converted to terraces for growing crops. Apart from this, significant decline in the forest area was not evident. But the quality of forest was severely damaged.

Forest in Begnas was an integral part of the farming system. Fuelwood for cooking and heating purpose, fodder for animals, leaf-litter for manuring, timber for house construction and for making agricultural implements and charcoal for the iron-works were derived from the forest. Even though the demand of forest product was higher[14] than that could be supplied from the natural growth of the forest in the village, the prudent way of managing forest and using forest-products slowed down the process of deforestation for many years. Plantation of fodder and fuelwood trees in upland and in common land was one way of reducing the pressure on the forest. But political factors and government's stewardship neither helped the underprivileged people nor improved the conservation of the forest.

When the study area was revisited in 1993, forest was again handed-over to people for the management and use. From casual observation it appeared that condition of forest was significantly improved in the past 6 years. But from informal discussions with villagers it also appeared that they were deeply concerned about the ongoing process of deforestation and they were about to do something. But the system of community management reintroduced in the village by the Government agency gave them a chance to show their commitment for the conservation of forest. In new management committees, altogether 8 in number, all the

---

[14]The annual wood requirement in Begnas VDC was estimated at 43,938 m^3 in 1985. But annual increment of 547 ha of forest was estimated at only 27,685 m^3 (Source: BTRT Project Document).

households were included and all, including the members of the occupational castes, were sharing equal responsibilities in protecting the forest. The distribution system of forest products, especially fuelwood, was different in different places. In some cases it was distributed based on family size but in most cases it was equally divided to all households. In some cases funds were created by the monthly donations of households to employ forest guard whereas in other cases farmers themselves looked after the forest on a rotational basis. Rules and regulations were also formulated to punish the persons bypassing the community regulations. But villagers were more concerned about difficulties in punishing the politically influential persons as well as those who were extremely poor.

D. Water resources and fishing

Begnas and Rupa lakes provided good opportunities for fish culture. This water resource was the common property and every citizens of the village had rights to the use of these lakes for fishing. But in the past, caste regulations kept villagers away from fish culture and only *pode*, whose traditional occupation was fishing, used these lakes for fishing and fish culture. In other words, about 30 *pode* families solely used these lakes. Even until 1987, fish culture was not considered as an important economic activity in the village and at that time the Fishery Development Office was encouraging *pode* families to practice cage culture for the production of fish in intensive manner. This office provided them the technical know-how and the Small Farmers' Development Project provided credit on a group guarantee basis taking the cage itself as the collateral for credit. The policy of the government was to allow only poor households to keep cage in the lake.

By 1993, families of higher castes including Brahmins and Chettris were also found to have farmed fish in these lakes. Most of them were small farmers who trampled the caste regulations and entered into fish farming because of economic necessity. Out of about 250 private cages in these two lakes[15], *pode* families owned only 16. Gurungs followed by Brahmins and Chettris owned more cages. Sixty households of Gurungs (also included households from outside the study area), 25 of Brahmins, 4 of Chettri and 22 of occupational castes had fish cages in the lakes (*pers. comm.* with Pitamber Koirala, JTA). Moreover, participation of Brahmins and Chettris had increased in fishing. They caught fish by throwing gill-net and then sold them in the market. In the process of fishing, these Brahmins and Chettris also came in close contact with *podes*. Due to this constant interaction, it was also reported that some relaxation was seen in the rules of commensality.

---

[15]At the time of revisit, all the cages of Begnas lake were taken to Rupa lake because the new dam constructed in the former lake increased the depth of water and thus the fresh water accumulated in the lake did not provide enough nutrients for fish. It was expecied that two more years might be required before the necessary fauna and flora could be developed in the lake.

It was not only poor Brahmins and Chettris but also the wealthier ones had entered into this fishing profession. But the latter had taken it as an opportunity for investment. They were not really involved in fishing activities because for this work they had employed *podes* as business partners. These traditionally skilled people looked after the technical matters of fish culture but the wealthier Brahmins and Chettris acted as the financiers. As the Bank then did not give loan without collateral for the establishment of enclosures for fish culture at a large scale, *podes* who were almost landless did not get such loans. But Brahmins and Chettris using their land for collateral obtained credit and then made associations with *podes* and ran the business of fish culture. This had also enhanced the cross-caste interrelationships. But whether such relationships would become permanent or remain transient depend much upon the benefit-sharing and work-sharing mechanisms. Most of the fish farming groups in Begnas did not run smoothly in the past mainly because of dispute in work-sharing mechanism. This was especially so if the group members were from heterogeneous background. Free-rider attitude and the habit of escapism were reported to be the main problems in running co-operative work in Begnas including fish-farming associations. But given the constant pressure from capitalists to allow them to establish large cages for the commercial production of fish, it might not take long time before politicians consent to their demand. In that case, cages of the small farmers would be evicted because they would not be able to compete. Moreover, because of large number of cages, the productive capacity of the lakes will decline. Then the status of *podes* and local people might change from that of business partners to laborers.

CHAPTER FOUR

FARMERS AND THE PROCESS AND TECHNIQUE OF THEIR DECISION-MAKING

The preceding Chapter, in essence, analyzed the decisions made by different kinds of farmers of Begnas in the utilization of their natural resources. This Chapter discusses how these decisions were made, i.e., the process and technique of decision making both at the family and the village levels. This process and technique of decision-making includes communication system within the family and in the village [for the identification of problems, setting objective(s) and finding alternatives], consultation and decision-making skills. These aspects of decision-making were also influenced by socio-psychological environments like farmers' goals and objective in farming (reasons for becoming farmers), optimism in life, attitude towards problems, risky choices and outsiders, beliefs in supernatural forces and ritual activities and their perceived importance of maintaining community way of life. In short, this discussion is required to understand farmers and their way of thinking. This chapter is, thus, divided into two main sections: socio-psychological context of decision-making and the process and technique of decision-making.

As discussed in the introduction, the general perception of farmers, especially younger people, is to move away from farming for non-farm job opportunities. Moreover, farming has reduced to a part-time activity for a large section of the farming population even though it is seen in the previous chapters that cropping intensity has increased in both upland and lowland. In these circumstances, it is important to understand farmers' perception, beliefs and other changes in behavior in order to design the intervention mechanism to bring about positive changes in farming.

A. Socio-psychological context of decision making

Reasons for becoming a farmer:

When farmers of Begnas were asked 'why they became farmers', the majority responded with the answer that their parents and forefathers were farmers and so they followed the same profession. In other words, they became farmers because farming was their tradition. The second important answer was 'lack of alternative employment'. Only few respondents reported that they were attracted to farming because of high financial gains. Not much variation was recorded in the ratings of reasons for becoming farmers across landholding and educational categories. The only exceptions were the landless households who reported that they came into farming mainly because they did not find other full-time job. These responses were obtained from the heads of households, who belonged to the older generation, which used to consider farming as the best occupation followed by business and service. But young people were found to consider the service as the best occupation. They had been in the farming only because they did not get non-farm jobs.

After the question about the reasons for becoming a farmer, respondents were asked to scale the extent of satisfaction derived from different attributes of farming in order to judge their goals and values in farming occupation. From this scaling, indices of satisfaction from different attributes of farming were developed. These attributes were grouped into two classes: financial and non-financial attributes.

On average, non-financial attributes of farming were found to be giving more satisfaction to farmers than the financial attributes (Table 4.2). The first three important attributes that gave maximum satisfaction were 'can earn more income, if hard work is done', 'can maintain medium standard of living', and 'can stay with family' (Table 4.3). These attributes were more related to security needs of life. The implication that can be derived is that farmers were more concerned with security goals in farming. The last three attributes, which gave less satisfaction were 'more income if risk is taken', 'expansion in the occupation', and 'innovative and progressive occupation'. This shows that farmers recognize higher risk in this occupation, and yet they prefer it as a traditional occupation over others. By farm size categories, landless and small farmers derived more satisfaction from financial attributes of farming, whereas medium and large farmers derived more satisfaction from non-financial attributes. The reasons may be that landless and small farmers were mainly concerned with income to fulfill physiological needs rather, whereas others were more concerned with other needs like freedom, or outdoor living, because most of them were also surplus producers. This variation in the attitude of different kinds of farmers was also reflected in their decisions as discussed in the preceding Chapter. For example, farmers with large land holding preferred quality or palatability rather than yield in selecting the rice varieties. The total average index of satisfaction was also higher among large and medium farm size groups than in the other groups. Brahmins and Chettries were more satisfied with both types of attributes of farming. This is mainly because of the correlation between landholding size and ethnicity.

Table 4.1: Farmers (in percentage) giving different 'reasons' for becoming a farmer (Begnas, 1987).

	- Farm-size categories -				
Reasons	Landless	Small	Medium	Large	Total
'traditional occupation'	40.5	45.7	45.2	48.2	45.5
'no other employment'	42.8	41.5	34.2	42.2	39.1
'more income in farming'	10.2	12.7	19.8	9.6	14.4

(Source: Household survey, 1987)

Table 4.2: Index of satisfaction with farming by farm-size (Begnas, 1987; Index ranges from 0 to 1, 0 indicates the minimum satisfaction and 1 the maximum satisfaction).

Farm-size categories	Financial Attributes	Non-financial Attributes
Landless	0.36	0.37
Small	0.39	0.34
Medium	0.48	0.57
Large	0.55	0.73
All households	0.45	0.49

(Source: Household Survey, 1987)

Above discussion leads us to the conclusion that, even in a village like Begnas, diversity exists as to the goals and values of farmers depending mainly upon the possession of resources. The existence of different interest groups, i.e., surplus producers having mainly non-financial goals in farming, and subsistence producers trying to exploit their own labor to increase production, in one and the same society was found limiting their decision making. Occasionally, this was also found leading to the conflicts, particularly in taking decisions at the village levels for the management of private as well as common resources. This becomes clear from the following response (in 1987) of an old farmer when answering the question, 'why they do not grow wheat in paddy land as winter crop by stopping animals from entering the fields and by using fertilizer'?

> "Large owners have more land on *phat* (paddy fields located in one location and having large area) and they are not interested in growing wheat. It takes time and cost to fence small and fragmented plots. Growing wheat in small pockets is also prone to bird damage."

But the situation was found to be changing when the study area was revisited after about six years. By 1993, the number of small and landless farmers had increased significantly. Farms of wealthier farmers were also divided among their inheritors and many surplus producers of the past were not able to produce sufficient food for their families. For them also, it was necessary to grow more crops in ***phat*** (paddy fields). Since the interest of many farmers was matched, the decision was taken at a gathering of villagers to impose restrictions on letting animals loose in *phat*, so that interested farmers could grow crops there. Since then, it became necessary to look after the animals and this eventually led to the decline in the *goth* system of manuring paddy land (see preceding Chapter). However, it does not imply the non-existence of conflict of interests in the study area. Farmers with more animals were hard hit by this decision. Despite the unsatisfaction, they still abide by the new regulation.

Table 4.3: Farmers' ranking of the attributes of farming in terms of satisfaction index in Begnas in 1986/87 (Index ranges from 0 to 1; 0 indicates the minimum satisfaction and 1 the maximum satisfaction)

Attributes	Type	Rank	Mean Index	S.D. of Index
'more income if hard work is done'	F	1	0.69	0.23
'modest living'	NF	2	0.58	0.26
'can stay with family'	NF	3	0.51	0.30
'more income than other occupation'	F	4	0.50	0.21
'freedom in job'	NF	5	0.48	0.29
'can do preferred work'	NF	6	0.47	0.30
'can continue tradition'	NF	6	0.47	0.31
'healthy and outdoor living'	NF	8	0.44	0.23
'regular income'	F	9	0.42	0.23
'more income if risk is taken'	F	10	0.42	0.22
'expansion in occupation'	F	11	0.34	0.20
'innovative and progressive occupation'	F	12	0.33	0.19

(F = Financial Attribute, NF = Non-financial Attribute, S. D. = Standard Deviation; Source: Household Survey, 1987)

Attitude towards land:

Land had been the principal resource in the village and its ownership was considered as one of the symbols of prestige. Farmers also attached ritual significance to land, which differed from one cultural group to another. Therefore, a survey of attitude towards land was important to understand the farmers' decision making in the use and management of land.

About half of the respondents accorded respect to landownership. They attached prestige to a person who had more land. Proportionately more small and landless farmers had such attitude, whereas large proportions of medium land holding farmers with high educational status felt otherwise. More than half of the respondents from low education level had such attitude, whereas less than one third of the respondents of high educational status reacted likewise (Appendix Table A.5).

Prestige on account of landownership came in two ways: firstly, by the power of land owners to mobilize tenants, and secondly, by the general practice of selecting them to act as village leaders as they could also give more time to community activities, if necessary, by employing laborers for their farm work. Even in recent times, most village leaders were from larger land owning families. Another important thing was that families who had enough land were found to be sending their children to schools in town. Among the village residents, most of the government jobholders were also from such families. Hence, villagers held such persons in high esteem.

To the question 'whether landowners' practice of letting land for share-cropping is a kind of help to tenants?', the majority of respondents, particularly the small and landless farmers, responded negatively. The reason for negative responses of small and landless farmers was due to the landowners' practice of giving distant and unproductive land for rental purposes.

Respondents' statements on the need for equal distribution of land were divided. Respondents giving positive or negative responses were almost equal in number. Variation in the response by different types of farmers was also found to be marginal. The major reason forwarded for equal distribution of land was related to food security for all the families. About one third of the respondents adopted that view. Other important reasons forwarded by those who favored equal distribution of land were related to increase in production and equal work opportunities. The reasons forwarded against the proposition were related to distribution according to capacity, luck or difficulty in having such egalitarian system.

Attitude towards leisure time:

As revealed from the discussions in the previous Chapter, use of family labor uniformly throughout the year was a major benefit of the diversified farming system. Similarly, farmers also experienced shortages of labor during peak farming season. Therefore, before the introduction of new technology, which demands more labor, it is imperative to see how motivated farmers are to sacrifice their leisure in order to increase their production or income. This also helps us to understand whether the farmers' adoption of traditional practices was due to their unwillingness to sacrifice the leisure time.

In the survey conducted in 1987, only about one fourth of the respondents reported that they were having enough leisure time, while others stated that they had to keep themselves busy throughout the whole year. The ratings of importance of leisure time showed quite a large variation among farm size groups. About one third of the landless respondents stressed that they attach much importance to leisure time, whereas more than half of the respondents from large land holding categories did report so. This was related to their goals in farming. As large landholders derived more satisfaction from the non-financial aspects of farming, they naturally tended to prefer leisure. More of the Brahmin and Chettri and educated respondents preferred to have more leisure time, the main reason for this being that they could produce food sufficient for their consumption.

The implication that may be derived is that to make an innovation adopted, it should not demand more labor time, as all the farmers are busy. It is particularly so for large landholders and higher educated respondents, as they attached more importance to having leisure time. But the strong majority of farmers were eager to sacrifice their leisure time, if they could increase their income significantly.

Optimism in life and of better productivity:

Optimism on the part of farmers motivates them to change their lives for better. Even the villagers had the saying that 'without optimism, new avenues of progress can not be found'. Despite the existence of such sayings, about one third of the respondents were convinced that their future life was uncertain, whereas one quarter of the respondents were of the view that their life would remain unchanged. Respondents optimistic of a better life were proportionately more frequent in numbers than those who were pessimistic of their prospects in life. Respondents from large land holding and higher educated backgrounds expressing optimism of a better life were proportionately more numerous than those from small landholding and less educated groups. More Brahmins and Chettries were optimistic than members of other castes, with a slight difference in proportions though. Regarding the possibility to increase productivity, more than half of the respondents had positive opinion.

Farmers' perceived problems:

Farmers' perception of the severity of various problems that they were facing was measured using the scaling method. The most urgent problems were related to land, followed by marketing and credit (Table 4.4). Land had been the major problem for all categories of farmers. Marketing problems were rated important by many members of the medium farm size group, whereas credit, input and labor problems were pronounced among the large farm size group.

Screening the individual problems (Appendix Table A.6), the most pressing problem was irrigation. This was true for all categories of farmers. This might be obvious in a rainfed area where uncertainty and inadequacy of rainfall are putting constraints on crop production. Other problems related to land, in their order of severity, were insufficient land, poor soil fertility, and fragmentation. Poor fertility was the result of surface erosion, and fragmentation was due to the inheritance system. Regarding labor problems, the most important issue was the high cost of labor. This had increased because of the availability of jobs in government sponsored development projects, where wages were generally high. Unavailability and high cost of fertilizers were the important problems related to inputs. High interest rate was named as a major problem related to bank credit as stated by members of small, medium and large farm size groups; whereas landless farmers rated the sheer unavailability of credit as their major problem. Regarding marketing, problems of pricing by middlemen were rated highest. These middlemen were mostly outsiders, with only few being residents of the village.

Despite existence of many problems, a majority of farmers were not deterred by problems. On the question of what they would do if they were deprived of various public facilities, about two thirds of the respondents reported that they would get to know the reason for it. About one quarter of the respondents were of submissive nature, as they said that `they would keep quite'. Only few of the respondents displayed an aggressive attitude, which was comparatively more frequent among the landless.

Table 4.4: Index of farmers' perceived problems in crop production by farm-size categories (Begnas, 1986/87; Index ranges from 0 to 1; 0 indicates the extent of problem at the minimum level and 1 the extent of problem at the maximum level).

Problems related to	Landless	Small	Medium	Large	All
Land	0.76	0.72	0.68	0.64	0.69
Labor	0.35	0.36	0.45	0.55	0.43
Input	0.49	0.49	0.52	0.56	0.51
Credit	0.49	0.48	0.54	0.56	0.52
Marketing	0.51	0.53	0.61	0.58	0.57

(Source: Household Survey, 1987).

Respondents' behavior toward natural problems was studied by means of their response to a hypothesized situation. More than half of the respondents were not deterred by such problems, stating that they would work as usual. About one-third of the respondents stated that they would work hard to compensate the loss. Migrating as a tendency in the fear of such problems was observed at very low frequency. Indeed, comparatively most respondents eager to migrate were landless. A hard working attitude in the face of such problems was displayed more frequently among the large farm size group (Appendix Table A.7).

The explanation that can be drawn from such response is that most landless and small farmers' fear of such problems. The probable reason may be that their possession of fewer assets, the principal asset being the land, aggravated by such problems may pose a threat to their lives. As already seen, almost three-quarters of the families had small land holdings that could not feed their families in a bad year.

Attitude towards bank credit:

Credit delivery mechanisms play an important role in increasing production by way of helping resource poor farmers to purchase inputs and increase their assets. But whether or not farmers take credit for these purposes also depend upon the prevailing perception about the credit. The survey of the household heads showed that slightly more than half of the respondents perceived of a risk in obtaining bank credit by placing their land as collateral. This risk feeling was more widespread among landless and small farm size groups and lowest among large farm size groups. The risk averse attitude of small farmers aroused from their fear of losing collateral due to the failure of the project undertaken with the help of credit from banks. It was also observed that small and underprivileged people were afraid of dealing with the government agencies.

This risk attitude was reflected in the appropriation of bank credit. Less than one fifth of the respondents had taken any bank credit. Despite the priority given to landless and small farmers in the bank's lending policy, proportionately more respondents from the medium and large farm-size categories were found appropriating bank credit[16]. Major enterprises for which bank loans were obtained included mainly livestock, fishery and purchase of fertilizer. Most of the loans taken by respondents with large farm size and

---

[16] Even though SFDP (Small Farmers Development Project) located at Begnas gave credit to small farmers, other farmers got loan from Agricultural Development Bank located at Gagangauda. It used to take about two hours on foot to reach to Gaugangauda from Begnas.

higher educational level, and belonging to the Brahmin or Chettri castes were spent on livestock and fertilizer, whereas others had taken loans mainly for fishery. Risk of losing the collateral due to failure of the enterprise was reported to be the main cause of not taking bank credit. This response was further validated from the fact that small and landless families had taken credit if they did not have to offer their property as collateral. For example, *pode* families received credit for fish culture because the cages bought from bank loan were considered as the collateral. Similarly, other small farmers were found to take credit on a group guarantee basis simply because they did not have to place their land as collateral.

Fatalistic attitude:

A fatalistic attitude is considered a constraint for efficient decision making. However, such beliefs were still existing in society, and some people took them into consideration while making decisions. Five common sayings expressing chiliastic expectations akin to fatalistic attitude were read to the respondents, to be scaled based on their beliefs in their message. The index thus prepared was used to measure this complex attitude.

The overall index was computed at 0.64, based on a maximum of 1.0. The virulence of this attitude was found to decrease as we proceed from lower to higher farm sizes and educational levels. This index was found to be lower among Brahmins and Chettris than among members of other castes (Table 4.5). Respondents held a strong belief in luck, which is thought to be provided by the supernatural power. The secondmost important belief was related to deeds in past lives. According to this belief, the present position of an individual is the result of his (her) deeds (*karma*) in the last birth and nothing can change their effects in the present life. Other beliefs were related to the effects of religious and merit making activities (Appendix Tables A.8 and A.9).

The explanation which may be derived from this analysis is that fatalistic attitude might have influenced farmers decision making in crop production. Small and less educated farmers' decision making might have been comparatively more influenced by such beliefs than members of the other groups.

Table 4.5: Index of chiliastic expectations akin to fatalistic attitudes by farm-size, educational level, and caste in Begnas in 1986/87 (Index ranges from 0 to 1; 0 indicates the minimum level of chiliastic expectation and 1 the maximum level of chiliastic expectation).

Farm-size	Mean Score	Education	Mean Score	Caste	Mean Score
Landless	0.77	Low	0.74	Brahmin & Chettri	0.60
Small	0.74	Medium	0.55	Others	0.72
Medium	0.58	Medium	0.55		
Large	0.50				

(Source: Household Survey, 1987).

Magico-astrological beliefs and merit making:

Farmers were found to be familiar with some magico-astrological beliefs, thinking that observing them improves.

productivity. However, some of them expressed skepticism of such beliefs.

An index of belief in magico-astrology was constructed based on the scaling of respondents' beliefs. Seven commonly held beliefs were identified and were given to villagers for scaling from which the index of magico-astrological beliefs was prepared. The index for the whole sample was calculated at 0.64 as against a maximum of 1.0 (Table 4.6). Strong beliefs were expressed for 'can not plough on new moon and full moon days' and 'women and Brahmins should not plough the land'. In others, respondents showed modest belief (Table 4.7). This index was high among the respondents with small farm size and low educational level.

It appears from the types of beliefs and ritual activities that farmers carry out these activities with an aim of solving problems, which are out of their control, like risks and uncertainties arising from the natural causes such as rainfall, hailstorm, earthquake and insects and pests. This is intimately linked to the worldview of people following the Hindu religion. According to this worldview, the world and its natural processes were developed and controlled by different types of Gods. To minimize the chances of occurring of a particular problem, the responsible God was appeased through particular ritual activity. Accordingly, the type of ritual activity to be organized depended upon the nature of God to be appeased. Many types of ritual activities were carried out every year on a community or a family basis. The ceremonies, which required elaborate activities, were organized only on a certain occasion depending upon the severity of the problem. Many of the resources like land, animal and forest were also considered sacred, and, thus, were worshipped every year. It is believed that this tradition was started with an aim of preserving these resources and avoiding their indiscriminate use.

Due to the existence of such beliefs, farming activities were guided not only by agronomic requirements of crops but also by the religious norms. For example, farming activity should be started on an auspicious day and farmers consulted local priest to find out such a day. It was believed that by starting work in such auspicious day, disastrous results could be avoided. As an example, Tuesday was considered best to start with the transplanting of rice. Farmers also performed religious rites with an aim of increasing production. A ceremony was organized on a community basis to beg for rain from Indra, the God of rain, at the start of the rice-planting season. After the rite, on the first Tuesday of 'Bhadra' (July/August), rice transplanting was prohibited. This time roughly coincides with a period after which rice growing becomes uneconomical due to complex reasons related to photo-period. Knowledgeable persons (local priests having knowledge of the astrology) were consulted to forecast the likelihood of rainfall, and accordingly farming operations were carried out. Some crops like sesame and barley, grasses like *kush*[17] and trees like *bel* (*Aegle marmelos*) were vital in religious ceremonies and members of the higher caste, especially the priests took special care to grow and maintain such crops and plants.

---

[17] Similar to Imperate grass, but leaf edges are sharper.

These practices sometimes posed constraints. Farmers might have been foregoing a rainy full moon or new moon day. This was more important during the crucial time of rice transplanting in rainfed areas such as in the study area. Small farmers were hard hit if this happens. Prohibition of women and Brahmins from ploughing became problematic during time of labor shortage. This was one of the reasons for the poor land preparation for paddy cultivation. Other practices did not pose such grave constraints. Ceremonial rites in the process of crop production might have hampered the work, but at the same time they provided some diversion and even entertainment to an otherwise tedious work. While forecasting was practiced, past experiences, were also taken into consideration. Thus, some authors pointed out that such forecasting is at least better than randomization. Regarding the utilization of past experience, one elderly man made the following prediction of hailstorm occurrence:

> "if there was a gap last year, chances are high that a hailstorm occurs during this year. It will occur in the same season as it occurred in the year before last. If a hailstorm occurred last year, it is less likely in this year, and if so it will occur in another season".

As discussed in the preceding Chapter, hailstorm have been causing more damage in villages of Nepal. Accordingly, many types of ritual activities were followed for the prevention of the occurrence of hailstorm. One common custom, which was still followed in villages, was that they employ Lama, practitioner of exorcism, who organized ritual ceremonies if he foresaw the chances of hailstorm, which was usually judged from the darkness of the cloud. Then, with his *trantric* (magic) power, it was believed, he concentrates the cloud in such a way that

Table 4.6: Index of belief in magico-astrology in Begnas in 1986/87 (Index ranges from 0 to 1; 0 is the minimum level of belief and 1 the maximum level of belief)

Magico-astrological statements	Mean (Index)	S.D. (Index)
'can not plough in new moon and full moon days'	0.88	0.95
'women should not plough the land'	0.87	0.26
'Brahmins should not plough the land'	0.60	0.32
'ceremonial rites in August protect crops from rats'	0.50	0.30
'more yield is obtained if the crop is planted on astronomically good day'	0.57	0.31
'*oli* can prevent hailstorm if he maintains purity'	0.51	0.33
TOTAL	0.64	0.21

(S.D. – Standard Deviation; Source: Household Survey, 1987).

Table 4.7: Index of beliefs in magico-astrology by farm size, education and Caste in Begnas in 1986/87 (Index ranges from 0 to 1; 0 is the minimum level of belief and 1 the maximum level of belief)

Farm-size categories	Mean Score	Education	Mean Score	Caste	Mean Score
Landless	0.76	Low	0.71	Brahmin/ Chettri	0.61
Small	0.72	Medium	0.57	Others	0.21
Medium	0.58	High	0.49		
Large	0.53				

(Source: Household Survey, 1987).

hailstorm drops on certain areas having no crops like barren land, forest or river. Such persons were locally called *oli* and every household paid their *oli* a certain amount of grain each year. Until recently, a Lama *oli* used to practice his exorcism in Begnas, but due to his increased negligence, many farmers stopped paying him and he no longer practiced exorcism related to hailstorm.

Social institutions and organizations:

Farmers of the study area belonged to different groups, had developed distinct relationships, and had specific beliefs and values. These institutions might have posed certain limits on crop production activities. As the natives of the village, farmers had many obligations like commitments to the community, membership in the co-operative labor group, irrigation society and school maintenance committee, liabilities as a borrower or lender, and obligations to guests or to outsiders. How these obligations put limits on the choice of crops or their varieties, or on production maximization, is discussed in this section.

i. Landlord tenant relationship:

Nature and types of relationship established between landlords and tenants affected decision-making on crop production. Landlords were found to give land for share-cropping to a tenant who had little land. They also gave preference to a tenant who belonged to their kin group. At the same time, they would evaluate the performance of a tenant in terms of the produce he delivered as their share. Thus, a landlord's decision on to whom to let land for crop cultivation was reached after considering the trade-off of these two implications.

In some cases, landlords were found to provide seed or animal power to their tenants. Sometimes, fertilizer cost was also shared, but in such cases tenants had to reciprocate by volunteering their labor. The selection of varieties was solely the decision of the landowners. If the land given for share-cropping was of excellent quality, landlords would want two-thirds of the produce. But generally total produce was shared equally between landlords and tenants. Landlords also influenced tenants' decisions by giving advice after their inspection of crops in the fields. Two farmers who rented in land on a fixed rental basis were given full responsibility for decision-making. Eight farmers who were cultivating *guthi* (temple land) were also independent in making their decisions.

The majority of the respondents stated that the present relationship between landlord and tenant was fair. About half of the respondents held the view that such relationship was good in the past, and the reasons given by most respondents were trust and helpfulness. The main reason forwarded by those respondents, who maintained that such relation was not good in the past, referred to landlords' exploitative behavior. Nearly three fourths of the respondents gave positive reactions to the need for maintaining a good relationship between landlord and tenant because they believed that this leads to the possibility of mutual benefit.

Thus, the decisions of tenants on crop production were not only determined by requirements imposed by plants and soil but also by those persons who owned the land. That obligation had the quality of an unilateral requirement dictated by landowners. But in recent times, the bargaining power of the tenants had also increased because of the availability of non-farm job in the town and in Government assisted projects. While production was still divided in the ratio of 50:50, the tenants demanded manure from the landlords.

ii. Obligations between neighbors:

Except for two individuals, all respondents stressed the necessity of a good neighbor, and the vast majority said that their neighbors had been good to them. In order to maintain a good neighborhood, certain obligations must have been fulfilled. Among many examples, the following were found having potential significance in the context of the study area with regard to crop production.

Where the lands of different farmers were adjacent, fencing was done on a co-operative basis, irrespective of whether one grew crops or not. But in most cases, farmers grew crops if they had a share in the maintenance of the fence. Damages by birds were very high when a crop was sown in a small field. Owing to this fact, farmers were forced to choose those crops, which were also grown by neighboring farmers. This was a constraint partly imposed by the small size of land holdings. Due to the same reason, the maturing date of different varieties, if sown together, should have been adjusted to more or less the same period. This was especially the case in rice farming. If all the neighboring paddy fields were harvested, farmers either had to protect their crop from animals individually or had to harvest prematurely, because of the tradition of letting the animals loose into the paddy fields after most had been harvested. But this custom was not found to have followed in many places when the study area was revisited in 1993.

Irrigation was an important activity where the neighborhood played an important role. In most of the cases, water was taken from springs to lower fields flowing through the terraces located at higher fields for rice cultivation under irrigation. If the rice in the upper field had just recently been planted, excessive water might have damaged the rice. As Farmer X said:

> "I had to replant rice in the places where seedlings were removed by excessive water, which was taken flowing through these terraces by some farmers, who have land at a lower place than mine, for their rice transplanting".

Farmer Y adjusted to the problem in a different way, as he stated:

> "I delayed the transplanting of rice in five *gara* (terraces) for ten days, until after the farmers, who take water through these *gara,* had finished their transplanting".

Farmers whose fields were located at the center were forced to transplant rice earlier to avoid damage caused to the surrounding fields of neighbors by the movement of oxen and people, as there were no trails or tracks leading to individual plots.

iii. Other obligations to society:

Some farmers had to act as office bearers in the many organizations, both formal and informal, existing in the village. Formal organizations at village level were the Village Panchayat (replaced by Village Development Committee in 1990), ward members council, four school maintenance committees and one pond maintenance committee. Other informal organizations at the village level, where farmers had to participate, were trail maintenance group, water channel maintenance group, forest conservation group, temple maintenance group, wedding group, and funeral group. Wedding and funeral works were taken more seriously than others, as they also were significantly related to crop production. At least one adult from each household, mostly among neighbors and relatives had to participate in such occasions, sometimes sacrificing one or two days of urgent work. Some religious ceremonies were organized at the village level every year and the number and types of ceremonies depended upon the village religious institutions like the number of temples, number and types of deity believed to exist in the village. It was also related to the cultural affiliation of the villagers. For example, in communities where members of Tibeto-Burman group resided, religious ceremonies were organized every year to worship forest and forest-protecting Gods. Every household of the village participated and contributed cash and kind donations in these ceremonies. But members of the occupational caste did not have to contribute in such ceremonies, mainly because they were considered religiously impure. Other obligations, generally, did not occur during the peak cropping season. The only festival, which coincided with the period of high labor demand, was *tihar*, which sometimes coincided with the harvesting of early paddy. In such cases, people celebrated *tihar* in the early morning hours and spent the rest of the time harvesting their crop.

From the respondents' high positive response on the need for social work and their willingness to participate in such kind of work, it may be concluded that farmers' decision making might have been affected by such commitments as well. Indeed, social work was organized during slack periods in major works.

There were 33 groups, as of 1987, which had taken bank credit on a group guarantee basis. But by 1993, the number of groups had increased to 57. Members of a group would hold meetings once a month, and a representative of the local bank would come to attend the meeting. They exert influence on the decisions farmers take. Farmer A was coerced by a bank employee to transplant the *mansuli* variety of rice in utilizing the money given in the form of credit for purchasing fertilizer. Farmers also offered best quality produce to such officials in anticipation of future benefits they hoped to avail from such persons.

Other social organizations directly related to decision-making in crop production were the co-operative exchange group, the iron-plow sharing system, and the ox sharing system. The iron-plow sharing system was more common at Begnas, where very few respondents kept this implement owing to small terraces. Because of its advantage in turning the soil and, thus, labor saving in preparing the land for millet transplanting, they developed such system. In forming the co-operative labor group, farmers gave priority to neighbors and kinsfolk because of ease of making arrangements. They had also the system of efficiency counting while exchanging labor, but at the same time also accepted younger people so as to provide them the opportunity of strengthening their efficiency. About half of the respondents arranged oxen on a share basis, which was more common among medium farm size groups. These social organizations had a lot to do with the timing of farming activities, which was decided on the basis of mutual agreements of group members. It was also said that in the past, households which finished their rice planting would donate one or two days of animal and human labor to households who, due to some problems, would not finish their rice transplanting in time. Despite the erosion in such systems, farmers donated labor to a person in distress. This was purely extended as a help without the expectation of any reciprocal return.

When the study area was revisited in 1993, more organizational groups were found to have formed, mainly because of the motivation from governmental and non-governmental organizations. Apart from 8 forest management committees, 19 soil conservation groups were formed. But in the past, there were more than 8 forest management groups and membership to these groups was based on caste or lineage. But the membership to the newly formed groups was based on the geographical location of the households rather than their economic or social class. As a result, these groups were much more heterogeneous than those in the past. Therefore conflict of interest for the sharing of work or for the distribution of benefits might have become a serious problem in these groups. But given the fact that caste restrictions were diluting and peoples' behavior was dictated more and more by economic necessity and that the economic mobility had increased, the conflict of interest might not take place strongly on caste line but on economic line[18]. Farmers were obliged to attend the meetings of these groups and made contributions in cash or kind.

The practice of treating guests and celebrating festivals had imposed a burden in the form of producing superior quality foods, because best quality foods were to be offered to guests, which were also required during festival time. Most respondents stressed that they treat a guest with the same honor as in the past. Due to this reason, the majority of farmers devoted some of their land to produce superior quality food grains in whatever small amounts even at high production cost. It was observed that such crops produced less yield in relation to the quality of land they required.

Some villagers had developed relationships with outsiders, particularly townspeople, with the expectation of some help when they had to go to the town for the purchase or sale, or for the work in Government office. When visiting

---

[18] After the reinstatement of democracy in 1990, ethnic consciousness has grown in Nepal. Accordingly, number of organizations with the aim of fostering ethnic and caste solidarity has increased since then. This is also true in the case of the study area, but at the same time inter-caste or inter-ethnic group interaction has also grown mainly due to economic necessity as discussed in Chapter 3.

the townspeople, villagers usually took special products of the village as gifts. In the household survey, three-fourths of the respondents agreed with the hypothesized case of developing a relationship with townspeople for the benefit of their help and knowledge of innovations. Eagerness to develop such relationship was slightly higher among households with large farm size.

B. Decision-making process and technique

The basic unit of the village society was the family. Accordingly, decision-making process at the grassroots level started from the family. But decisions were also made at the village level, which, as discussed above, also affected the decision-making at the family level.

The process of decision-making at the family level was also similar to the concept of people's participation which means involving people in planning (identification of problems and available opportunities), decision-making, implementation and evaluation. These steps were also taken in a family environment, but in a very informal way.

The decision-making at the family started with the need to solve the existing problem in the family or in the farm, or to solve the problem related to the adoption of new ideas and exploration of new opportunities. After this, all the family members were consulted and when certain conclusions were derived by the consensus, the head of the family, usually the male, announced the decision. Before announcing the decision, the family head made sure that the right decision had been made based on right information. The withdrawal of a decision after its announcement would adversely affect the social standing of the family, and neighbors and fellow villagers would start to suspect the integrity of the family. Members of the higher castes were particularly sensitive to this because, in the village, they were considered behaving like a member of a low caste if they did not stick to their word. Due to this custom, the head of a family always tried his best to make a right decision by consulting his family members and close associates in the village. Family members would not fully co-operate in implementing a decision taken without their consent.

One of the reasons for consulting all family members in making decisions was that farm work was also divided based on the age and gender[19] The problem in a particular area was more clearly understood by the family member undertaking that particular work. For example, problems about the animal feed (fodder and grass) and firewood were more clearly understood by the women and they could also offer viable solutions. Similarly, problems about tending animals were well known to the children. Since the changes in one aspect of farming affected all other aspects of family life, suggestions of all members needed to be taken into account by the family head. It was due to this reason that even the ploughman - which actually did not belong to the family, particularly in the case of Brahmin and Chettri

---

[19] A generalized model of division of work by age and gender is presented in the Appendix Table A.11.

families - was consulted and his suggestions were taken into account.

Division of work in a family showed that women undertook a large share of the family and farm work. They also played a greater role in farm decision making not by announcing the decision and taking responsibility for it, but by providing information to the decision-maker. The wife (in few cases wives) of the decision-maker played an important role in channeling the information to her husband[20] because daughter(s)-in-law who usually shared a large burden of the farm and family work, did not directly communicate with the father-in-law. But it was generally the case that mother-in-law tried her best to exploit her daughter(s)-in-law and, thus, they were usually not in good terms. In that case, the latter conveyed their feelings through their husbands. If the conflict became intense, it paved the way to the separation of married children from the main house and then they started taking independent decisions about their family and farm. Friction within the family was less intense and less frequent in families having no married son(s). The wife was not expected to criticize her husband and his decisions in public places. This was especially so in the case of Brahmin and Chettri families. Similarly, division of work by gender was

[20]In most of the families males were the decision-makers. Only in few cases, women were found to take decisions for the family; this happened if the husband was not at home because of their temporary migration or the son(s) were not matured enough to take the responsibility. Elder son generally took the responsibility if he was matured enough to do so. A study conducted in 1992 in this village revealed that in 78.6 per cent households, household heads were men. Women were household heads only in 21.4 per cent households (Pandey, 1992).

followed more strictly in their cases, whereas considerable extent of flexibility was observed among families of Gurungs, Magars and occupation castes. In their cases, women would take decisions ignoring their husbands, but very rarely this would happen in a Brahmin family. Among Thakalis, the managerial function of the household or their hotels and restaurants was undertaken generally by the women.

The above discussion about the decision-making within the family was based on the experience and the observation of the author. A survey conducted for this study also contained the questions about the consultation of the decision-maker within and outside the family. The survey showed that almost all respondents consulted their family members while taking decisions related to crop production. The few respondents who did not consult their family members were mostly single or had a small family. Not much variation was found in this regard by different categories of respondents.

Ranking of family members on the basis of value attributed to their suggestion showed that the spouse was ranked first, children second and parents third. The reason was that parents, in most of the cases being very old and, hence, no longer acting as household heads and respondents, were not supposed to know much about the present cropping situation. However, following the tradition in itself was a consultation not only with old people of this generation but also with the people of past generations. It was also a custom in the village to seek advice from older people while taking a crucial decision.

Outside their families, respondents attributed highest value to the suggestion of neighbors, followed by relatives. The formal leaders, namely the representative of the ward to the village council, head of the village (which was called Pradhan Pancha in 1987) and JTA[21] were ranked third, fourth and fifth respectively. Respondents' reaction by farm size and caste were consistent with the whole sample.

Regarding the frequency of consultation, respondents ranked neighbors, relatives, ward member, JTA and village head in descending order. This order shows that social distance was important for the frequency of consultation. Neighbors and relatives usually included members of the kin group. People generally went out of their kin group for consultation only if they were very close friends or had good relationships. But for the collection of information, they did not limit, as discussed in the following section, themselves within the kin group or the village. By farm size, medium farmers were found to be frequently consulting the JTA, only next to neighbors, whereas others placed the JTA in the third rank.

The decision-making process at the village level was also the same as described for the family. Village council generally kept a crier who announced by shouting the time, place and agendas of the proposed meeting. But since

[21] JTA (Junior Technical Assistance) was the government employee working at the village level for the dissemination of improved farming know-how.

household heads participated in such meetings, they tended to be dominated by the males. Decision was generally reached by the consensus, but group pressure was also commonly applied to persuade those who did not conform to the majority.

Government's efforts to promote agricultural development had helped in the creation of communication network between villagers and outside agencies. Male members of the family, particularly the household head, introduced new ideas and information to the family members, as they were generally the ones to go out. Respondents were also asked about the sources of information about the farming technology and about their habit of experimenting. This is covered in the following sections.

i. Sources of information:

The major source of information for villagers about agricultural technology, as reported by respondents, was the radio-broadcasting program, followed by neighbors and relatives (Table 4.8). Other important sources were local leaders and posters as well as pamphlets. Neighbors and relatives were the main sources of information for landless and small farmers, whereas that for medium and large farmers was the radio program. Local leaders were also important sources of information for landless and small farmers. For respondents of low education levels, the main sources of information were neighbors and relatives, whereas for those of medium and high education levels it was the radio program.

Table 4.8: Sources of information (by farm size) about modern farming techniques used by farmers of Begnas in 1986/87.

Sources	Landless	Small	Medium	Large	All
Neighbors and relatives	35 (36.1)	72 (35.5)	76 (29.8)	39 (27.3)	222 (31.8)
Local leaders	17 (17.5)	27 (13.2)	33 (12.9)	11 (7.7)	88 (12.6)
JTA	6 (6.6)	10 (4.9)	20 (7.8)	17 (11.9)	53 (7.3)
Traders/salesmen	7 (7.2)	14 (6.9)	12 (4.7)	3 (2.1)	36 (5.1)
Radio	28 (28.9)	67 (32.8)	81 (31.8)	47 (32.9)	223 (31.9)
Posters and Pamphlets	4 (4.1)	14 (6.9)	33 (12.9)	26 (18.2)	77 (11.0)

(Source: Household Survey, 1987).

Villagers also went to government offices to collect information, however, only one-third of the respondent said that they had been to such offices. Proportionately, more respondents from large farm size and high education level groups were found to have visited government office. The mean number of visits during the year prior to this survey by those who had visited these offices was 3.5. The majority of those who did not visit these offices reasoned that they did not expect to receive any useful suggestion. Distance was found to be the second most frequent reason for their reluctance (Appendix Table A.10).

Villagers were found to have the habit of talking about crops, yield, cost and other similar matters whenever they had a chance to gather in some public places. During these discussions, they evaluated several possible alternatives pursued by different farmers. The systems of caste or kin group were not found inhibiting the flow of information, as upper caste and lower caste people were found to be practicing free exchange of ideas.

ii. Individual experiment:

More than two thirds of the respondents said that they would experiment first before adopting whatever kind of innovation. The remaining one-third stated that they would not wish to experiment. The habit of experimenting was least frequent among the landless farmers, and its frequency increased as the land size increased. The less frequent habit of experimenting among small farmers was certainly due to the risk involved in this exercise and the lack of resources for this purpose. Even the larger landholders did not grow new crops or varieties at a large scale. At first they would plant them in a corner of a terraced field or in a small terrace. After observing its growth and yield, they would make their decision on adoption or rejection. Farmers also observed any experiment done by others in the village. Farmers knew the benefit of trial cultivation. The majority of them stressed the importance of getting to know the suitability of any new crop or variety. When the first author revisited the field in 1993, he noticed village residents talking about the farming of a conservation farmer, officially recognized as an extension agent at the grassroots level. This farmer had adopted different types of crops and fruit trees in his small landholding. He was officially considered as the model for organic farming to be imitated by others. But other farmers were of the opinion that the model farmer was not able to make any significant profit from the new cropping pattern. They were also startled by the decline in his economic position. They were also seemed to take note of every cash/kind transaction in the model farm. They said that they would wait and see how much profit he could make from his fruit trees and coffee before taking any decisions on the adoption of some of his practices.

Farmers, as a result of their experimenting habit, had adopted a variety of crops, mostly fruit and grasses. Citrus fruits like orange, mandarin and lemon were grown in the village. Fodder grass and trees were also well adopted. In fact, those crops and methods were adopted if farmers found them suitable in their fields and were profitable in terms of providing food or some cash income.

Decision-making skill and technique:

Farmers in the study area were not found to keep records of plans and accounts of their farming activities. Even though, they were conscious of the cost and benefits of various enterprises within their farms, they were not in a position to tell the financial or the economic standing of their farm. Lately, the practice of cultivating cash crops had also started mainly because of increased demand for cash income. Farmers knew more about the financial returns of cash crops than that of other crops, which were generally meant for consumption at home.

Considering the average education score of a family, it could be said that as a family, farmers possessed sufficient skills to plan for and keep account of their cropping activities. The only thing that was found lacking was the formal training on such planning and account keeping. However, about three-fifths of the respondents pointed out that they made plans mentally of their activities before they started working their fields. The reason given by most of those farmers, who did not maintain any records, was that farming being a small business they did not see the necessity to do so. Others were of the opinion that record and account keeping was not necessary in farming as it was carried out with an aim of producing food for home consumption.

Respondents who reacted positively to the need for account keeping reasoned that such practice helped to know the costs and benefits of crops as well as their shortcoming and, thus, provided some orientation towards proper investment. The advantages of prior planning, as they felt, were related to success of their cropping enterprise, ease of work, and savings on time.

Farmers had learnt about the farming practices through informal training from their elders. Children also learnt about farming by observing their parents working in the fields. Thus, an integrated and organized socialization process, centered around agricultural production, made children to become farmers. The strong majority of the respondents believed that their children could learn about farming from their parents, however, slightly less than half of them cautioned that such skills learnt casually were not sufficient to face new challenges imposed by the small size of land holdings.

The implication is that farmers, despite their knowledge about planning and account keeping, did not use these methods of management in farming. Some of them, indeed, were found to have kept records if the need arose. Thus, the majority of the farmers did not apply any business skills to farming.

It was also difficult for farmers to run farms purely on business line because social and moral ideology came in the way of maximizing family benefits. For example, farmers decision-making as seen from practical examples presented in Chapter 3 was also influenced by the objective of enhancing prestige or, at least, not damaging one's social standing in the village for which one should be bounded by the caste and village-centered morality. On the other hand, wealth had been increasingly recognized as a symbol of prestige for which family-centered rationality needed to be pursued. Due to this reason, some wealthier villagers practiced various unfair means to concentrate wealth like charging exorbitant interest rates. The issue of prestige was also based on the norms and values developed in the village. For example, taking credit was taken as a symbol of one's worsening financial position. If any one in the village were found to be taking credit, then he/she would become a subject of gossip. It was also due this fear that some farmers went to informal sources for credit (moneylenders) and usually fell into the trap of vicious circle of indebtedness. Similarly, high expenses in ceremonies like marriage were considered to boost one's standing in the society. Accordingly, some families organized such ceremonies even at the cost of their land.

It was also commonly practiced in village environment that people hide the truth or tell only the half-truth about a problem on which a decision was going to be taken. One main reason for this was that there had been internal rivalries between families and they competed each other to take new opportunities or advantages, be that in farming, politics or in employment. It is because of this reason that the matter to be decided was kept secret before taking any decision and only trusted and close friends and kinsfolk were consulted. This was particularly true in the case of strategically important decisions. As intrigue and deception were not uncommon in village society, these methods were employed to maximize the interest mainly at the family level.

It appears from above discussion that social life in village Nepal was a mixed-blessing of harmony and conflict. Harmony had been maintained, with the help of village level organizations and institutions, to preserve the village life and village resources, whereas conflicts had become intense because of villagers' decision-making with a point of view of family-centered rationality to maximize income or prestige.

CHAPTER FIVE

RATIONALITY IN FARMERS' DECISION-MAKING

Chapters 3 and 4 discussed the opportunities and limitations offered by the physical and socio-cultural environments in making decisions about the farming activities. Discussions in Chapter 3 clearly showed that farmers had been able to exploit the advantages of diversity in the physical environment. This Chapter also showed that the farming system they had developed was oriented towards their main strategy in farming, the survival of the family. Chapter 4 demonstrated that socio-cultural environment, though appeared in some cases to have inhibited the farmers to maximize their income at the family level, did not put, in most of the cases, any constraints on their (farmers) adoption of appropriate modern farming practices. On the other hand, the perception of farmers was to reduce risks to their livelihoods. The belief system was also oriented towards maintaining a survival of villagers, even though in some instances it also inhibited profit maximization at the individual level. The discussion in these Chapters was not geared to objectively determine the effects of various factors related to these aspects (physical, cultural and socio-economic environments) of the society on the decisions made by farmers. Therefore, an attempt has been made in this Chapter to identify various qualitative and quantitative factors that were discussed in the preceding Chapters, and to analyze their effects on the adoption of improved technology of rice production. Rice was selected for this purpose because it was the staple food in Nepal and in the study area. Here, adoption of appropriate modern technology has been assumed as the criteria for rational decision-making in farming, because it is mainly through the adoption of such technology that improvement in agricultural production and in the economic conditions of farmers can be achieved.

As discussed in Chapter 1, it is commonly argued that the influences of socio-cultural and institutional factors on farmers, which also affect their beliefs, perceptions and attitudes, do not permit them to adopt efficient and productive activities in place of traditional ones. It is due to this reason that farmers in developing countries are not considered rational in their decision-making. But the main argument of this thesis is that though socio-cultural and institutional factors influence farmers lives, they do not pose any significant constraint in the adoption of modern technology if it is appropriate. The hypothesis is precisely formulated as follows:

Hypothesis:

> **Farmers' decision on the adoption of modern technology is dependent upon many types of factors. Indeed, environmental factors play an important role in the adoption process. Socio-cultural factors, as such, do not explain variation in adoption. Thus, it is hypothesized that farmers' beliefs and values do not pose any significant constraints for farmers' decisions on the modernization of farming.**

The multivariate analysis was the method employed to test the hypothesis and this analysis included many variables (factors) on farmers' beliefs, ritualistic and other cultural behaviors. These variables were developed by allowing respondents to rate their beliefs on a scale. Indices were then developed based on scaling of each variable. Farmers were categorized based on these indices. These variables were then included in the analysis to examine their effects on the adoption of modern technology in rice production.

In the context of the study area, dissemination of modern technology was mainly concerned with the improved varieties and chemical fertilizers. This study had considered the use of *mansuli* variety of rice and all types of chemical fertilizer as the adoption of modern technology. The *mansuli* was the variety considered as high yielding and also recognized as the suitable variety in the hilly region. The combination of these two innovations was considered important because they were equally essential, and they should also be adopted simultaneously in order to increase production and improve productivity.

The following critical variables were taken into consideration representing different types of factors as stated in the hypothesis: total annual rice production, household size, respondent's age, respondent's education, composite family education level, farm size, non-farm occupation, optimism of better productivity, practice of planning, practice of selecting varieties on the basis of high yield, contact with townspeople, exposure to outside media, availing of institutional credit, visit to extension office, ritualistic attitude, magico-astrological belief, contact with local leaders, contact with traders and salesmen, willingness to take risks, accessibility to market, optimism in life, awareness of modern technology, family labor force ratio, and irrigated land holding.

Quantification Theory II, which is similar to discriminant analysis, was used to test this hypothesis. Partial correlation of Quantification Theory is comparable to the coefficient of dependent variable in the regression analysis. After making appropriate categories of each variable, a preliminary analysis was done using Chi-square test between predictable and each critical variable.

The correlation ratio of 0.534 shows that 53.4 per cent of the variation in the adoption of modern technology could be explained through these variables. Partial correlation showed the accessibility to market as the important variable. The secondmost variable was willingness to take risks. Optimism of better productivity and availing of institutional credit were found negatively related to acceptability of modern technology. The ranking of the variables on the basis of partial correlation is given in Table 5.1.

The effect of composite family education level was found to be higher than that of respondent's education. Other important factors affecting the adoption in a positive way were magico-astrological beliefs and contact with local leaders.

Table 5.1: **Ranking of variables on the basis of partial correlation (the result of QUANT II analysis).**

Variables	Categories	Partial Correlation	Rank
-'Yearly total rice production'	3	0.173	8
-'Household size'	3	0.174	6
-'Respondent's age'	3	0.069	17
-'Respondent's education'	4	0.144	9
-'Family education'	3	0.174	6
-'Farm size'	4	0.078	15
-'Non-farm occupation'	2	0.077	16
-'Optimism of better productivity'	2	-0.054	24
-'Practice of planning'	2	0.107	12
-'Practice of selecting varieties on high yield basis'	2	0.016	21
-'Contact with townspeople'	2	0.057	20
-'Exposure to outside media' (2 criteria)	2	0.110	11
-'Availing of institutional credit'	2	-0.015	23
-'Visit to extension office'	2	0.066	18
-'Fatalistic attitude' (5 criteria)	3	0.099	13
-'Magico-astrological beliefs (7 criteria)	3	0.228	3
-'Contact with local leaders'	2	0.183	4
-'Contact with traders/salesmen'	2	0.179	5
-'Willingness to take risk'	2	0.291	2
-'Accessibility to market'	2	0.411	1
-'Optimism in life'	3	0.086	14
-'Awareness of modern technology' (5 criteria)	3	0.066	18
-'Family labor ratio'	3	0.135	10
-'Irrigated land holding'	3	0.000	22

Correlation ratio = 0.534
Square root = 0.731

Accessibility to market was the main factor that explained much of the variation in adoption. Thus, environmental factors played an important role in the adoption of modern technology. Cultural, social and psychological factors did not explain much of the variation in the adoption process. Given accessibility, combined with contact with traders and salesmen, being important in adoption, it stands to reason that availability of modern inputs (technology) and farmers' relationships with the outside world (cosmopolitan) played a major role in accelerating the adoption process.

This analysis, thus, falsify the commonly held opinion that peasants beliefs, attitudes and their socio-cultural environment have been impeding them to move towards modernization in farming. Accordingly, there are no grounds to label peasants' decision-making as irrational on account of their beliefs, attitudes and socio-cultural environment.

Impressions gained from the revisit to the study area in 1993, after about six years of the previous fieldwork, also suggested that farmers show considerable flexibility in following religious and cultural norms while adopting appropriate technology which was accessible to them. Involvement of members of higher castes in the business of fish production was a clear example in this regard. Similarly, traditional system, which had been impeding the maximization of production, was also found to have changed. For example, the tradition of letting animals loose in the paddy fields (*khet*) after the harvest of the main season paddy had also been abandoned. By 1993 three crops were grown in a large proportion of *khet*, whereas just about 6 years ago, only a single crop of paddy was grown in such field. Improved varieties of maize like Manakamana and Arun-2 were well adopted in the village. Use of chemical fertilizer, pesticides and insecticides had significantly increased and it had also become necessary to use these inputs to maintain the production of crops. Regarding the role of village organizations and institutions, it was indeed seen that they put many obligations on individual farmer. Obligations to kinsfolk and neighbors and to the village society in some cases were found to have inhibited individuals to maximize the benefits. But these organizations had also been helping in the conservation and utilization of common resources like forest and irrigation system, which were also intimately integrated with the utilization of private resources like land and livestock. Moreover the institutions like co-operative groups had been helping people to perform agricultural works in efficient manners. It is therefore, argued here that farmers' decision-making was rational and at the same time aimed at securing the survival of the family. Accordingly, any innovation, which brought risks to the survival of people, had not been adopted.

CHAPTER SIX

CONCLUSIONS AND RECOMMENDATIONS

A. Decisions for survival

The management of subsistence oriented farming system of Begnas was guided mainly by the tradition, but changes were also introduced whenever the appropriate technology was available. Changes in the production system were also necessary to increase the production due mainly to the economic necessity, the result of growing population pressure and increased demand for cash income. Despite the changes, the fundamental orientation of the traditional farming, i.e., diversification and integration, had been retained, mainly because of its suitability to the physical environment of the village. Moreover this system had many other advantages which augmented the food security at the household level.

Farming was also intimately integrated with the family and village life of the farmers. Accordingly, the social organization of the family and the village and cultural and religious inclinations of various groups of farmers had also been affecting farmers' decision-making. But the beliefs, perception and attitudes of farmers did not impede the adoption of modern technology. In fact, it was revealed from various evidence gathered from this study that innovations that were suitable to the village environment and having beneficial effects on the whole farming system were successfully adopted.

Resources (physical environment) and farming:

Farmers were found to use resources located at different places in an integrated way. The farming system they had followed was thus of the diversified and integrated one. Suitability of upland and lowland for different crops with different seasonal requirements in labor input helped farmers to use the family labor more uniformly throughout the year. For example, paddy used to be grown in lowland and maize and millet in upland. These crops did not compete each other for the use of farmers' time. But if farmers had grown a single crop in upland\ or lowland, problem of labor shortage would have been great. Labor relation within the village was also found to have eased the problems related to labor supply. There wee certain agricultural tasks which, once started, needed to be finished as quickly as possible. To cope with this requirement, farmers had developed the system of exchanging labor and forming co-operative groups.

The diversified farming system also had many other advantages like production of a variety of foods for home consumption and low level of risk and uncertainties arising from natural causes like hailstorm, rainfall, insects, pest and disease. Loss from storage would also be less as crops were continuously grown and used.

Location of fields (upland or lowland) in relation to houses and forest was found to be affecting the utilization of these fields. For example, upland land was located near the houses as they were built at the top portion of the hill flank. This location was not only good for healthy living, but also near the forest. As a result, less time was required for the collection of firewood, tree-fodder and leaf litter. Since, animals were kept near the house, collected manure was easy to use in upland than carrying down the hill. Accordingly, collected manure was used in upland. However, in upland, individual fields were fenced by stone walls or by bamboo, which permitted greater individual freedom for growing crops. But, in lowland, such system did not exist and crop cultivation was more or less determined by the joint decisions of households (i.e. by society). But by 1993, after 6 years of the previous research, a considerable increase in the intensity of crop production in lowland was also observed. Farmers brought this change because of their economic necessity, mainly the more demand of food due to population growth. This change also led to considerable social rearrangements.

Livestock, forests and lakes were other resources in the village. Like land, livestock was also owned on a private basis, but forest and lakes on a common basis. The farming system followed by households integrated both private and common resources. It was due this reason that village level decisions affected the decision-making at the household level.

Utilization of common resources like forest and lake-water in recent past demonstrated farmers' flexibility in their adherence to religious norms. When faced with survival crisis in the farms, some farmers (especially small and marginal farmers) moved towards the utilization of income opportunities in fish farming, which, in the past, was permissible only to *pode* caste due to caste restrictions. Moreover, fish farming had also been a good field for investment for the wealthier farmers. Because of involvement in new enterprises like fish farming and changes in the management systems of local forests, new inter-caste or inter-ethnic relationships were found have evolved. The traditional patron-client relationships between high caste and low caste groups had slowly moved towards the direction of relationships of business-partners. As a result, discrimination against members of low caste was found to have declined, but not completely removed.

Farmers and the process and technique of decision-making:

The socio-cultural and socio-psychological environments were found to have affected farmers' way of thinking and their decision-making. Therefore, a slight difference among cultural groups in their way of making decisions was observed. Moreover, farmers were also differed in their goals and objectives, beliefs and attitudes according to their land holding and education. The organization of the society and of the family also influenced the decision-making process, i.e., flow of information, communication, consultation, decision-taking, implementation, evaluation and taking responsibility for the decisions.

Farmers' decision-making was obviously affected by their goals, values, beliefs and attitudes to problems and by their way of solving the problems. Different categories of farmers differed in these aspects because of the differences in their possession of resources and cultural background. There was significant variation in the indices of satisfaction with financial and non-financial attributes of farming. Farmers were found to be deriving more satisfaction from non-financial attributes of farming. Small farmers' satisfaction index for financial attributes was significantly higher than for non-financial attributes. This variation was an outcome of their differing needs. Small farmers were concerned with fulfilling their basic minimum needs. Hence, the majority of them found farming a risky and non-innovative occupation. Accordingly, risk and uncertainties affected farmers' decisions. Farmers were also found to have developed mechanisms to cope with uncertainties that arise in farming, mainly due to natural causes. Small and medium farmers were found more sensitive to risk and uncertainties because of their fewer resources. Risk-averse attitude of small farmers was also reflected in their borrowing bank credit by placing their assets as collateral.

As the status of a large proportion of households had lowered to that of small farmers because of division of farms among inheritors (mainly sons), they felt the pains of risks and uncertainties. As a result they preferred a secure source of income. Accordingly, service in government or non-government sector had become the priority of many village youths. On the other hand, even those having sufficient land to feed the family (locally called landlords) preferred to have job because it brought not only economic progress, but also social prestige. Farmers of this class also preferred to have leisure time, which could be utilized for social and political purposes.

Fatalistic attitudes stemming from religious beliefs or chiliastic expectations were found to be deeply-rooted in the minds of the people, as documented through the index of fatalistic attitudes which was estimated at 0.64. A significant difference was observed in the fatalistic attitudes among different age, education, and size of landholding groups. Farmers also held certain magico-astrological beliefs. The estimated index of these attitudes was quite high at 0.64. Some of these beliefs posed constraints on crop production, while the general adherence to specific patterns had some regulative advantages.

It seemed from the type of ritual activities performed by villagers that these activities were concerned with the farmers' way of solving problems, which were out of their control. Minimizing risks and uncertainties from natural causes like rainfall, hailstorm, insect, pest and disease were the prime aims in organizing ritual activities in relation to farming.

Various types of social institutions existed in villages were found to have put some limit to as well as provide some opportunities for increasing production. These institutions entailed some obligations on the part of farm households. Indeed, these obligations sometimes led to more production on a sustainable basis as most of them were related to water management, cropping schedule and selection of crops and varieties. Farmers were also found to have abandoned the institutions not conducive for the adoption of appropriate farming practices. Instead, they had developed new institutions suitable to the changed circumstances.

Decision-making process at the grassroots level started from the family. Despite different roles and status of members of a family in villages of Nepal, all members were consulted before making decisions. The division of work by age and gender also made it necessary to consult all members of the family. But due to hierarchical organization of the family, some members were not able to pass information to the decision-maker directly and freely. Particularly, female members of the family were not given opportunity to make decisions, even though information was sought from them. Accordingly, in a large proportion of households males were the household heads, who also represented the household for village level discussions or activities. As a result, males also dominated decision-making at the village level.

Farm families did not apply business skills in the sense that they did not keep any records of works or the account of expenditure and income. But they were found to be conscious of relative costs and benefits in different farming enterprises. As reported by farmers, their main concern in farming is to produce various types of foods for family needs. The issue of prestige (caste and village centered morality) and family-centered rationality including the economic necessity were found to have guided farmers in making decisions. Villagers believed that prestige could be earned through *dharma* (merit making) and by following caste regulations and village customs strictly. But they were often in a conflicting situation whether to choose prestige or the family- centered rationality, i.e., maximizing benefits at the family level, or to abide by village and caste centered morality. In recent times, economic necessity was found to have forced a large proportion of farmers to increase benefits for the family even at the cost of prestige, i.e., not following the caste or village regulations. Again these decisions were made for the survival of the families.

B. Policy recommendations

The findings of this research have policy implications with regard to the development and dissemination of technology, commercialization of farming and to the people's participation in developmental activities.

Technology development and dissemination:

The existing diversified and integrated farming system was most advantageous to farmers as it enabled them to utilize the diverse ecological environment in the best way. This system also had many other advantages oriented at securing livelihood of the family. Therefore, improved technology should be developed in such a way that the fundamental elements of this system remain intact. For this, a whole package of innovations related to all components - agricultural production, livestock raising and forestry - of this system is required to develop, i.e., an integrated approach to agricultural system development is necessary. Livestock raising in the study area being not sufficiently productive, improved breeds of livestock along with proper support material need to be popularized. In addition to community forestry that is being promoted in recent times, system of private forests in many of the less productive land may be an effective approach to increase the forest cover in the village and to increase the income of the farmers. Integration of lowland cropping and upland cropping within a farming system should also be encouraged because of different timings of labor requirement for wet rice and dry upland crops. Similarly, diversity in the physical and socio-cultural environment also means that a blanket recommendation for a certain technology is not suitable. At the same time, technologies requiring more labor inputs may not be accepted as farmers are experiencing the shortage of labor during peak farming seasons because of schooling of children and withdrawal of village labor in non-farm employment.

As revealed from this study, a majority of farmers were small farmers who operated farming in an average landholding of half a hectare. This small land holding was also fragmented due to the system of land inheritance. At present, their livelihood was partly supported by the existence of common resources like forest and pasture. Therefore, common resources should also be preserved for the livelihood of these small farmers. The technology should be developed to suit the small land holdings these farmers have. Improved breed of goat and cattle is suitable for them, as these animals do not require much land. Similarly cultivation of high value crops like vegetables (for seed or for vegetable) in intensive manner in a part of their land many help them to generate cash income locally.

Since the purpose of farming of all farmers was to meet the food requirements of the family, new techniques, which pose even a slight risk to a family's food supply, should not be recommended. This is especially so in the case of small farmers. Farmers' risk taking capability should be built up, either by providing them with productive assets or through insurance, before technologies involving individual risk are introduced. For long-run solutions, however, an appropriate program of agrarian reform should be devised to increase the assets of these people.

Education should be given first priority in development programs. Education enables people to avail themselves of equal opportunity. Educational attainment also enables people to find non-farm jobs, which will contribute to relieving the pressure on the land and slowing down fragmentation. Even though, educated people, at present, were not showing interest in farming, they would be attracted in this profession once modern infrastructures like irrigation and road are available. Adoption of modern technology will also be accelerated after this stage of development. Not many problems will surround the implementation of this proposition, as respondents were self-motivated to have their children receive formal and better education.

For the dissemination of the farming technology, extension agents should give attention to the way farmers make decisions regarding farming operations. Farmers' were of the habit of consulting with people who were close to them. Accordingly, a suitable communication strategy ought to be devised to spread information, recognizing the two-step flow of communication involving mass media, first, and local opinion leaders, secondly. In this regard, an improved strategy must entail the purposeful involvement of expert farmers with leadership potential.

This research demonstrated that farmers' behavior was rational by most of the selected criteria. The system they were following was geared to optimizing their preferences, which include mainly the survival guarantee. They were also prepared to change their habits and traditions, if and when the need arises. Thus, it should be borne in mind by policy formulators that the important thing is not to doubt and speculate whether farmers might accept new techniques or not, but to devise and/or promote appropriate technologies.

Commercialization of farming and modern technology:

The focus of government intervention (for example the recent Agriculture Perspective Plan - APP) is the commercialization of farming for increasing cash income. On the other hand, the approach of farmers has been to retain farming as a subsistence-oriented enterprise (i.e., to meet family requirements and generate a small amount of cash income locally) and earn cash income to meet family requirements from the non-farm employment, mainly the labor migration. As the contribution of former in meeting the family's food requirements has been declining, farmers are now focussing on the later.

Considering the fact that commercialization of farming would expose the farmers to the vagaries of the market[22], it is not seen to improve the food security of small farmers. These small farmers dominate the rural population in Nepal. Therefore, for their interest also, it is pertinent that the subsistence orientation of existing farming system (as people themselves are doing) should be retained.

The technology so far introduced and to be introduced under APP relates to green revolution technology, which involves heavy use of high-energy inputs backed by sound research and extension systems. At present, most of the technologies and inputs that the government agents recommend to farmers are imported from foreign countries. Even though the use of these inputs had enabled, as revealed from this study, farmers to increase production and intensify farming, various side effects were also observed. These side effects included, among others, decline in fertility of soil and incidence of insect, pest and diseases. But the government agencies were not in a position to make constant innovations in technology to counteract these negative consequences. This had inflicted a considerable pain on farmers, as they did not know the remedy of emerging problems. Government agents also did not know the solutions to some new problems as they lacked resources to conduct research. Accordingly, farmers have remained depended on research undertaken in

[22] Failure of market mechanism leading to occasional drastic increase in prices making poor and small farmers unable to purchase the food in the market. Similarly, occasional drastic reduction in prices of produce of the farmers would ruin them.

other parts of the world, where their local problems may not be considered important. Therefore, before the introduction of any new innovations, government agencies should also consider their ability to conduct further research to make innovations so that the new problems are solved. Otherwise, the constant dependence on other countries for innovations may become costly, especially in future when all types of subsidies (after the full implementation of World Trade Organization's agreement) in agriculture will have to be removed. The case of fertilizers is a case in point. Until the late 1980s, the prices of fertilizers were kept low through subsidies from government, which also motivated farmers to use them. This had also been helpful in intensifying cropping pattern and to increase total food production. But after the implementation of liberalization policy, especially since the early 1990s, the price of fertilizers soared rapidly. Farmers were then in a dilemma: using fertilizers means requirement of heavy amount of cash income, but on the other hand, fertilizers were a must to maintain the production. This led to the use only Urea (nitrogen fertilizer), which was relatively cheaper and was thought to show quick response. But it had severe negative consequences on soil fertility and food quality. Considering this reality, it is necessary to think before hand the possibility of providing support for research and subsidy on a long-term basis. Accordingly, only those technologies and inputs should be promoted for which necessary supports can be provided on a longer-term basis.

People's participation:

For encouraging people to participate in developmental programs, the decision-making process of such programs should resemble the one that is followed at the family and village levels. The leader or the external agent should consult all and take decisions based on the consensus or the large majority. It is also necessary to present the participation in developmental work as a matter of prestige. Most households would participate in a work that is going to enhance their prestige even if it means a small cost to them. But to solicit the people's participation in a sustained manner, it is imperative to understand the ability and willingness of different households or groups to participate. For example, large landholder may not participate in implementing a project if it is going to demand labor because they value their leisure time. They would rather donate cash for such purpose. Moreover, they may take active part in making decisions about the goals and objective of the programs. As they occupy most of the leadership positions in the political and other institutions, they play a major role in directing the project in their interest. Their contact with people working in Government bureaucracy also helps them to lobby for their interest. On the other hand, poor people may not be able to take part in the discussions, which might lead to the finalization of project goals because they have to be busy in earning their livelihood. Therefore, it is also necessary to organize poor people into groups and to involve the representative members of groups in a project aimed at their welfare. Moreover such organizations help them to work in a co-operative way sharing and exchanging resources.

Through this way, their position in the local power structure will also be strengthened.

The above-discussed strategy for increasing the participation of rural poor is also well applicable in the case of women. Relying heavily on the village political institutions, women's participation can not be achieved as it is seen that head of the households, usually the male, take part in such institutions. But there are also informal organizations in villages like co-operative labor groups solely participated by women. These organizations may be strengthened or new organizations formed which become entry points for external agents to involve local women in developmental projects.

GLOSSARY

abal	Best fertile soils as classified under the local system.
ansi	Crescent shaped tools used for cutting purposes.
bari	Fields where crops other than rice are grown.
bauso	Medium size implement used for digging purpose.
bhal choupa	Rainfed paddy fields.
char	Soils poor in fertility as classified under the local system.
doyam	Soils good in fertility as classified under the local system.
goth	Temporary animal shed built on the fallow paddy land for cattle grazing.
guthi	A type of tenurial practice under which public lands are allowed to be used for crop cultivation for the benefit of certain religious institution.
hal	Amount of land area that is ploughed by one pair of oxen in one day's work.
khet	Field where rice is grown.
khukuri	Implement used for cutting purpose; its shape is slightly curved.
kodalo	Implement used for digging purpose; it is bigger in size than other digging implements.
kuto	Implement used for digging purpose; it is smaller in size than other digging implements.
panchayat	Assembly of people's representatives. It was also a political system (from 1960 to 1990) in which political parties were banned.
phat	Rice fields of one location.
sim	Soils fair in fertility as classified under the local system.
thimaha	A variety of crops which deviates from its original characteristics.
tihar	A Hindu and Buddhist festival of light, with illumination of houses and buildings.

Equivalents

1 *ropani*	0.05 ha
1 *muri* maize	63.20 kg
1 *muri* millet	72.64 kg
1 *muri* wheat	63.20 kg
1 *muri* mustard	59.80 kg
1 *muri* rice	48.80 kg (in unhusked form)
1 *muri*	20 *pathi*; 1 *pathi* = 8 *mana*
1 US $	N Rs 49. 50 (in August 1993) and N Rs 68.15 in July 25, 1999)

Abbreviations

SFDP	Small Farmers' Development Project
JTA	Junior Technical Assistant
FDO	Fishery Development Office
ALU	Average Livestock Unit
MCI	Multi-Cropping Index
LUI	Land Utilization Index
VDC	Village Development Committee
GDP	Gross Domestic Product
APP	Agriculture Perspective Plan

REFERENCES

Adhikari, Jagannath, 1999: "Remittance economy of Nepal" *The Kathmandu Post*, June, 23. (Daily newspaper published from Kathmandu).

Abarintos, E.P. et al, 1975: "Decision Making and HYV of Rice" *Journal of Agricultural Economics and Development*. 5(2): 139.

APROSC and JMA, 1995: Nepal : "Agriculture Perspective Plan". Kathmandu, Agricultural Project Services Center and John Mellor Associates, Inc.

Alviar, N.G. et al., 1978: "Decision Making of Rice Farmers on Fertilizer Use" *Journal of Agricultural Economics and Developement*. 8(1): 34.

Anderson, J.R., J.L. Dillion and J.B. Hardeker, 1979: "Introduction to Decision Analysis" in *Agricultural Decision Analysis*. J.R. Anderson, J.L. Dillion and J.B. Hardeker (eds), Ames, Iowa: Iowa University Press, p. 12.

Anderson, J.R., J.L. Dillion and J.B. Hardaker, 1979: *Agricultural Decision Analysis*, Ames, Iowa: Iowa State University Press. p. 10.

Arkin, Herbert and Raymond R. Colton, 1963: *Tables for Statisticians*, New York: Barnes and Noble (Second Edition).

Asian Development Bank (ADB) and HMG/N, 1884: "Strategy for Agricultural Development in Nepal", Kathmandu: Ministry of Agriculture, His Majesty's Government of Nepal.

Berry, Sara S., 1984: "Households, Decision Making and Rural Development: Do We Have To Know More?" Cambridge, Mass.: Harvard Institute for International Development, Harvard University, p. 1-2.

Binswanger, H.P., 1978: "Risk Attitudes of Rural Households in Semi-Arid Tropical India" Economic Growth Center, Discussion Paper No. 275, New Haven: Yale University.

Buddhaphitak, Soparth, 1980: "Analysis of the Adoption of Agricultural Innovations: A Case Study in Chiangmai Province, Northern Thailand". Bangkok: Asian Institute of Technology (AIT Thesis, HS 1980-3).

Calavan, Michal, 1977: *Decisions Against Nature: An Anthropological Study of Agriculture in Northern Thailand*, DeKalb, Ju: Northern Illinois University, Center for Southeast Asian Studies.

Chayanov, A.V., 1966: *The Theory of Peasant Economy*, Daniel Throner (ed.), Basile Kerblang and R.E.F. Smith, Homewood, Ill: The American Economic Association.

Dasgupta, Sipra, 1970: *Agriculture, Producer's Rationality and Technical Change*, Bombay: Asian Publishing House, pp. 36-61.

Demaine, Harvey, 1987: "Magic and Management: Methods of Ensuring Water Supplies for Agriculture in South East Asia" *Nature and Man in South East Asia*, P.A. Stott (ed.), London: School of Oriental and African Studies, pp. 49-67.

Dillion, John L. "The Definitions of Farm Management" *Journal of Agricultural Economics*, 3(2): p. 257.

DFAMS, 1984: "Agricultural Statistics of Nepal, 1983". Kathmandu: Department of Food and Agricultural Marketing Services, His Majesty's Government of Nepal.

Eicher, Carl and Lawrence Wilt, 1964: *Agriculture in Economic Development*, New York: McGraw-Hill.

Epstein, T.S., 1962: *Economic Development and Social Change in South India*, Manchester: Manchester University Press.

Evans, Grant, 1988: "The accursed problem : communists and peasants" *Peasant Studies*, 15(2) : 73-102.

Fallon, Edward Bernard, 1983: "The Peasants of Isan: Social and Economic Transition in Northeast Thailand", Madison, University of Wisconsin (Unpublished Ph.D. Thesis).

Firth, Raymond, 1965: "Social Structure and Peasant Economy: The Influence of Social Structure upon Peasant Economy" *Subsistence Agriculture and Economic Development*, C.R. Wharton Jr. (ed.), Chicago: Aldine, p. 27.

Foster, George M., 1962: *Traditional Cultures and the Impact of Technological Change*, New York: Harper.

Gasson, Ruth, 1973: "Goals and Values of Farmers" *Journal of Agricultural Economics*. 24(3): 521-537.

Gurung, Chandra P., 1981: "Peoples' Participation in Local Level Planning: A Case Study from Nepal". Bangkok: Asian Institute of Technology (AIT Thesis, HS 81-8).

Hagan, R. Albert, 1971: "Farm Planning at Microlevel", Kathmandu: Singh Durbar Press, His Majesty's Government of Nepal, p. 1.

Harrison, Mark, 1984: "Chayanov's Theory of Peasant Economy" *Rural Development: Theories of Peasant Economy and Agrarian Change*, John Harris (ed.), London: Hutchinson, p. 246 ("Hutchinson University Library").

Heady, Earl. O, and John L. Dillion, 1969: *Agricultural Production Function*, Ames Iowa: Iowa State University Press, pp. 108-110.

Heyer, J., 1972: "An Analysis of Peasant Farm Production Under Condition of Uncertainty" *Journal of Agricultural Economics*, 23(2): 135.

Hudson, N. W., 1980: "Social, Political and Economic Aspects of Soil Conservation" *Soil Conservation : Problems and Prospects*. R.D.C. Morgan (ed.), Chichester: John Wiley, pp. 45-54.

Johnston, R., Gregory Derek and Smith, David M., 1986: *The Dictionary of Human Geography*. Oxford: Blackwell Reference.

Joshi, Sri K., 1981: "Peoples' Needs, Bureaucrats' and Decision Makers' Preferences: A Case Study of Khopasi Village Panchayat", Bangkok: Asian Institute of Technology, (AIT Thesis HS 81-12).

Kay, R. D., 1981: *Farm Management, Planning, Control and Implementation*, New York: McGraw-Hill, p. 9.

Kroeber, A.L., 1948: *Anthropology*, New York: Harcourt, Brace and World.

Lado, Cleophas, 1988: "Farmers' Identification of Agrarian and Environmental Problems and Some Management Strategies in Maridi District, Southern Sudan" *Singapore Journal of Tropical Geography*, 9(1): 45-59.

Lipton, M., 1968: "The Theory of the Optimizing Peasants" *The Journal of Development Studies*, 4(3): 327-351.

Long, Normam, 1982: *Introduction to the Sociology of Rural Development*, Suffolk: English Language Book Society.

Low, A.R.C., 1974: "Decision Taking Under Uncertainty: A Linear Programming Model of Peasant Farmer Behaviour" *Journal of Agricultural Economics*, 25(3): 311.

Malla, Parasar B., 1983: "Logit Analysis of Technology Adoption by Rice Farmers in Dhanusa District, Nepal". Kathmandu: Agricultural Development Council, APROSC, p. 1 ("Research Paper Series" No. 22).

Manandhar, Sangeeta, 1982: "Farmers' Decision Making on the Production of Their Major Crops", Bangkok: Asian Institute of Technology, (AIT Thesis, HS 1982-14).

McPherson, Malcolm, 1986: Why Do Researchers Continue to Ignore Risks in Tests of Farmers Efficiency? A Comment on Shapiro's Rejection of The "Efficient But Poor" Hypothesis" *The Journal of Development Studies*, 22(3): 604-607.

Moerman, Michael, 1968: *Agricultural Change and Peasant Choice in a Thai Village*, Berkeley, Cal.: University of California Press.

Mosher, A.T., 1965: "The Development Problem of Subsistence Farmers: A Preliminary Review" *Subsistence Agriculture and Economic Development*, Clifton R. Wharton Jr. (ed.), Chicago: Aldine.

Mumford, J.D., 1981: "Pest Control Decision Making: Sugar Beet in England" *Journal of Agricultural Economics*, 32(1): 31.

National Planning Commission (NPC), 1970: "Fourth Five Year Plan (1970-75)", Kathmandu: National Planning Commission, His Majesty's Government of Nepal, p. 1.

Nash, Manning, 1971: "Market and Indian Peasant Economies" *Peasant and Peasant Societies*, Teodor Shanin (ed.), Harmondsworth: Penguin, pp. 158-178.

Norton, G.A., 1976: "Analysis of Decision Making in Crop Protection" *Agro-ecosystems*, 3(1): 27-44.

Norton, G.A. and J.D. Mumford, 1982: "Information Gaps in Pest Management" *Proceedings of the International Conference on Plant Protection in the Tropics*, p. 589.

Norton, G.A. and J.D. Muford, 1982: "Decision Making in Pest Control" *Advanced Applied Biology*, 8: 87.

Ortiz, S., 1967: "The Structure of Decision Making" *Themes in Economic Anthropology*, R. Firth (ed.), London: Tavistok.

Pandey, Ganesh Bahadur, 1992: "Gender Analysis of Environmental Conservation in Upland Agricultural Systems : A Case Study in the Middle Hills of Nepal", Bangkok: M.Sc. Thesis, Asian Institute of Technology (AIT HS 92-21).

Polany, Karl, 1957: "The Economy as Instituted Process" *Trades and Markets in the Early Empires*, K. Polany, C.W. Arensburg and H.W. Parsons (eds), Glencoe, Ill: The Free Press.

Popkin, Samuel L., 1979: *The Rational Peasant. The Political Economy of Rural Society in Vietnam*, Berkeley, Cal.: California University press.

Poudyal, Lokendra Prasad, 1990: "People's Involvement in Planned District Development Through Decentralization in Nepal", Bangkok: Asian Institute of Technology, Unpublished PhD Thesis.

Powell, J.D., 1970: "Peasant Society and Clientilist Politics" *American Political Science Review*, 54(June): 411-425.

Redfield, Robert, 1955: *The Little Community*, Chicago: University of Chicago Press.

Redfield, Robert, 1960: *Peasant Society and Culture*, Chicago: University of Chicago Press.

Rogers, Everett M., 1965: "Motivations, Values and Attitudes of Peasantry" *Subsistence Agriculture and Economic Development*, C.F. Wharton Jr. (ed.), Chicago: Aldine, p. 115.

Ruthenberg, Hans, 1980: *Farming Systems in the Tropics*, Oxford: Clarendon, (2nd Edition).

Schultz, T.W., 1964: *Transforming Traditional Agriculture*, New Haven and London: Yale University Press, p. 37.

Scott, James C., 1977: *The Moral Economy of Peasants: Rebellion and Subsistence in Southeast Asia*, New Haven and London: Yale University Press.

Sen, A. K., 1981: *Poverty and Famines. An Essay on Entitlement and Deprivation.* Oxford: Oxford University Press.

Segura-de los Angeles, Marian, 1986: "Upland Economics and Economic Impact Analysis" *Man, Agriculture and the Tropical Forest: Change and Development in the Phillipine Upland*, Sam Fujisaka, Percy E. Sajise and Romulo del Castillo (eds.), Bangkok: Winrock International, pp. 169-188.

Shanin, Teodor, 1984: "Polarization and Cyclical Mobility: The Present Debate over the Differentiation of Peasantry" *Rural Development: Theories of Peasant Economy and Agrarian Change*, John Harries (ed.), London: Hutchinson, p. 223 ("Hutchinson University Library").

Shanin, Teodor, 1971: "Peasantry as a Political Factor" *Peasants and Peasant Societies*, Teodor Shanin (ed.), Harmondsworth: Penguin, pp. 238-264.

Shanin, Teodor, 1971: "A Russian Peasant Household at the Turn of the Century" *Peasants and Peasant Societies*, Teodor Shanin (ed.), Harmondsworth: Penguin, p. 31.

Shanin, Teodor, 1971: *The Awkward Class : Political Sociology of the Peasantry in a Developing Society : Russia 1910-1925.* Oxford: Clarendon Press.

Shapiro, K.M., 1983: "Efficiency Differentials in Peasant Agriculture and Their Implications for Development Policies" *Journal of Development Studies*, 19(2): 179-190.

Sharma Ramesh P., 1980: "Uncertainty and Subjective Beliefs in the Adoption of Modern Farm Techniques: A Case Study of Nepalese Farmers", Kathmandu, Agricultural Development Council, APROSC, 1980, pp. 1-2, (Research Paper Series No. 5).

Shiva, Vandana, 1991: *Violence of the Green Revolution : Third World Agriculture, Ecology and Politics.* Penang: Third World Network.

Shrestha, B.P., 1981: *An Introduction to the Nepalese Economy,* Kathmandu: Ratna Pustak Bhandar, p.1.

Sie, Kwat Soen, 1968: "Prospects of Agricultural Development in Indonesia (with reference to Java)", Wageningen: Center for Agricultural Publishing and Documentation, pp. 5-7.

Singh, Stavir and J.J. Roy, 1980: "Modernization and Development among Indian Farmers. A Modern Proof of Some Old Theories" *Economic Development and Cultural Change*, 28(2): 509-522.

Stirling, Paul, 1971: "A Turkish Village" *Peasants and Peasant Societies*, Teodor Shanin (ed.), Harmondsworth: Penguin, pp. 37-48.

Swaminathan, M.S., 1982: *Science and Integrated Rural Development*, New Delhi: Concept Publishing House.

Tantray, Abdul Majid "Adoption of Agricultural Innovations in Kashmir Valley, Northern India", Bangkok: Asian Institute of Technology, (AIT Thesis, HS 81-23).

Tiwari, Bishwa Nath, 1991: "Stagnant Nepalese Agriculture: Agrarian Reform or Environmental Management" *Nepalese Economic Review*, 5(1): 21-32.

Uphoff, Norman, 1978: "Problems Inhibiting Achievement of Boarder People's Participation" *People's Participation in Rural Development in Nepal,* Kathmandu: APROSC, pp. 86-71.

Upton, Martin, 1973: *Farm Management in Africa: The Principles of Production and Planning*, London: Oxford University press.

Wahab, A.M.F., 1979: "An Analysis of Factors Influencing Adoption and Non-adoption of High Yielding Varieties of Rice in Bangladesh" Bangkok: Asian Institute of Technology, 1979 (AIT Thesis, HS 79-2).

Waterson, A., 1965: *Development Planning*, Md.: John Hopkins University Press.

Watts, M., 1983: *Silence Violence : Food, Famines and Peasantry in Northern Nigeria*. Berkeley: California University Press.

Wharton, Clifton R. Jr., 1965a: "The Issues and A Research Agenda" *Subsistence Agriculture and Economic Development*, Clifton R. Wharton Jr. (ed.), Chicago: Aldine, pp. 455-467.

Wharton, Clifton R. Jr., 1965b: "Subsistence Agriculture: Concepts and Scope" *Subsistence Agriculture and Economic Development*, Clifton R. Warton Jr. (ed.), Chicago: Aldine, pp. 12-22.

Wharton, C.R., 1965c: "Subsistence Agriculture, Concept and Scope" *Subsistence Agriculture and Economic Development*, C.R. Wharton Jr. (ed.), Chicago: Aldine.

Wolf, Eric R., 1966: *Peasants*, Englewood Cliffs, N.J.: Prentice Hall.

World Bank, 1983: *Development Report*, Washington D.C.: World Bank.

APPENDIX

Table A.3: Month-wise labor input on an average farm (in mandays) in Begnas, 1986/87.

Months	Wet Rice	Dry Rice	Maize	Millet (R)	Millet (S)	Wheat	Total
May	4.5	0.0	10.1	0.0	0.0	0.4	4.6
June	5.4	1.9	5.3	0.0	0.0	0.0	3.8
July	36.0	1.9	1.0	0.9	0.0	0.0	15.2
August	1.8	11.2	5.3	0.6	0.4	0.0	4.2
September	18.9	0.7	0.0	12.3	0.4	0.0	10.0
October	1.8	3.5	0.0	1.8	8.6	0.0	10.3
November	18.9	0.6	0.0	0.6	1.3	0.0	9.2
December	18.9	4.8	0.0	0.6	0.4	0.0	9.0
January	0.0	0.0	0.0	6.0	4.4	0.0	5.4
February	4.5	1.6	0.0	1.5	1.1	1.3	4.8
March	0.0	0.0	2.4	0.0	0.0	0.4	1.1
April	0.0	0.0	2.9	0.0	0.0	0.8	1.6
S.D.	11.37	3.22	3.19	3.64	2.58	0.42	10.99
Mean	9.92	2.18	2.24	2.02	1.38	0.24	17.35
C.V.	114.6	147.7	142.4	180.2	186.9	175.0	63.30

Note : Only major activities in crop production were considered. Millet (R) – Millet relay cropped with maize, Millet (S) – Millet sequentially cropped with maize; S.D. – Standard Deviation, C.V. – Coefficient of Variation. Source: Own survey, 1987.

Table 2: Important local varieties of main crops grown in Begnas in 1986/87.

Rice : Thima, Thakali, Ramani, Darmali, Badana, Sindure, Gurdi, Mansara, Godi, Pahale, Jhinuwa, Anadi, Kathegurdi, Jathobudo, Marsi, Jarneli, Sijali, Anga.

Maize : Sano-seto (small-white), Sano (small), Thulo (big).

Millet : Kartike, Mangsire, Dalle, Agaute, Ashoje, Khalse, Asare.
(Source: Own Survey, 1987)

Table A.3: Approximate losses (in %age) caused by hailstorm in rice and maize crops during the years 1977 to 1985.

Individuals reporting Losses	Rice					Maize		
	1977	1979	1982	1985	1986	1978	1981	1985
1	20	33	50	25	15	33	60	10
2	25	40	45	30	25	30	75	5
3	20	33	40	33	20	25	65	7
4	25	30	45	30	15	30	70	10
5	25	30	40	33	20	20	75	5
6	20	40	45	30	15	25	55	5
7	20	40	50	33	15	30	65	5
8	25	35	45	25	15	30	75	10
9	20	33	45	25	20	35	65	10
10	20	33	40	33	15	30	65	15
11	30	35	40	35	10	25	60	10
12	25	35	45	35	15	20	65	5
13	15	33	50	33	15	20	75	10
14	20	35	40	33	10	25	55	5
15	25	30	45	35	10	30	60	10
16	20	35	45	35	10	35	60	5
17	15	35	40	30	15	20	66	10
18	25	40	50	33	20	25	66	15

(Soutce: Own survey, 1987)

Table A.4: Index of satisfaction with farming by farm size in Begnas, 1986/87 [Index range from 0 (minimum satisfaction) to 1 (maximum satisfaction)].

Attributes	Landless	Small	Medium	Large
A. Financial Attributes (total score)	**0.43**	**0.47**	**0.58**	**0.67**
- regular income	0.30	0.36	0.47	0.57
- more income if risk is taken	0.30	0.33	0.47	0.53
- more income if hard work is done	0.58	0.65	0.70	0.81
- expanding occupation	0.26	0.27	0.36	0.47
- innovative and progressive occupation	0.27	0.25	0.35	0.43
- more income than from other occupation	0.43	0.47	0.54	0.49
B. Non-financial Attributes (total score)	**0.33**	**0.41**	**0.67**	**0.87**
- freedom in job	0.24	0.29	0.59	0.71
- modest living	0.39	0.50	0.63	0.77
- can do liked work	0.23	0.26	0.56	0.77
- healthy/outdoor living	0.24	0.26	0.53	0.73
- can stay with family	0.28	0.35	0.56	0.79
- can continue the tradition	0.29	0.38	0.55	0.59
Total Average	**0.33**	**0.41**	**0.67**	**0.87**

(Source: Household Survey, 1987)

Table A.5: Prestige of land ownership and interest in having more land, by farm and education level [number of respondents] (Begnas, 1986/87).

Respondents	Prestige on land ownership?		Desire for more land?	
By farm size	**Yes**	**No**	**Yes**	**No**
Landless	27 (69.2)	12 (30.8)	35 (89.7)	3 (7.7)
Small	47 (59.5)	29 (36.7)	64 (81.0)	15 (19.0)
Medium	35 (37.6)	56 (60.2)	57 (61.3)	36 (38.7)
Large	26 (47.3)	28 (50.9)	25 (45.5)	30 (54.5)
By education				
Low	75 (55.2)	57 (41.9)	102 (75.0)	33 (24.3)
Medium	54 (49.1)	55 (50.0)	68 (61.8)	42 (38.2)
High	6 (31.6)	12 (63.2)	11 (57.9)	8 (42.1)
All households	**135 (50.8)**	**125 (47.0)**	**181 (68.0)**	**84 (31.6)**

(Source : Household Survey, 1987).

Table A.6: Index of farmers' perceived problems by farm size in Begnas, 1986/87 [Index ranges from 0 (minimum level of problem) to 1 (maximum level of problem)].

Problems	Landless	Small	Medium	Large	All
Land (Average)	0.76	0.72	0.68	0.64	0.69
Land shortage	0.92	0.74	0.66	0.50	0.69
Low fertility	0.70	0.70	0.65	0.62	0.67
Fragmentation	0.54	0.57	0.59	0.63	0.95
Lack of irrigation	0.87	0.84	0.85	0.79	0.84
Labor (Average)	0.35	0.36	0.45	0.55	0.43
High charge	0.45	0.42	0.49	0.64	0.50
Unavailability	0.33	0.40	0.48	0.61	0.46
Poor performance	0.30	0.29	0.38	0.45	0.36
Lack of skills	0.32	0.36	0.47	0.52	0.43
Input (Average)	0.49	0.49	0.52	0.56	0.51
Unavailability	0.46	0.52	0.55	0.55	0.53
High price	0.46	0.49	0.52	0.54	0.51
Lack of know-how of application	0.53	0.41	0.50	0.53	0.49
Credit (Average)	0.49	0.48	0.54	0.56	0.52
Unavailability	0.61	0.51	0.57	0.55	0.56
High interest rates	0.59	0.56	0.63	0.65	0.61
Bank at distance	0.37	0.39	0.43	0.43	0.41
Difficult procedure	0.45	0.48	0.58	0.59	0.55
Marketing (Average)	0.51	0.53	0.61	0.58	0.57
Lack of market	0.44	0.50	0.53	0.61	0.58
Exploitative pricing by middlemen	0.49	0.56	0.65	0.55	0.58
Traders cheating on weight	0.45	0.54	0.59	0.51	0.54
Low price level	0.60	0.57	0.74	0.75	0.67

(Source : Household Survey, 1987).

Table A.7: Respondents' reactions to natural hazards by farm size and educational level [respondents and percentages] (Begnas, 1986/87).

Respondents	Work as usual	Migrate	Work harder	Others
By Farm size				
Landless	21 (53.8)	3 (7.7)	9 (23.1)	6 (15.4)
Small	47 (59.5)	1 (1.3)	25 (31.6)	6 (7.6)
Medium	50 (53.7)	2 (2.1)	35 (37.6)	6 (6.4)
Large	28 (51.0)	0 (0.0)	24 (43.6)	3 (5.4)
By education				
Low	75 (55.1)	5 (3.7)	44 (32.3)	12 (8.8)
Medium	64 (58.2)	0 (0.0)	39 (35.4)	7 (6.4)
High	7 (36.8)	1 (5.3)	9 (47.4)	2 (10.5)
By caste				
Brahmins and Chettris	100 (54.3)	3 (1.6)	69 (37.5)	12 (6.5)
Others	46 (56.1)	3 (3.6)	24 (29.3)	9 (11.0)
Total	141 (54.9)	6 (2.3)	93 (35.0)	21 (7.9)

(Source: Household Survey, 1987) – Figures in brackets are percentages.

Table A.8: Index of fatalistic attitude by farm size [Index ranges from 0 (minimum level of fatalism) to 1 (maximum level of fatalism] in Begnas in 1986/87].

Reaction to the following statements	Landless	Small	Medium	Large
- 'present bad condition is the result of wrong deeds in the past life'	0.87	0.87	0.65	0.62
- 'what is written in the forehead (or fate) is never written off'	0.90	0.92	0.76	0.63
- 'one day God will give a happy day to all'	0.74	0.64	0.49	0.42
- 'present problems are created by God'	0.71	0.67	0.48	0.42
- 'religious work solves the present problems'	0.70	0.67	0.50	0.47
Total	0.77	0.74	0.58	0.50

(Source : Household Survey, 1987).

Table A.9: Index of fatalistic attitude by education level [Index ranges from 0 (minimum level of fatalism) to 1 (maximum level of fatalism] in Begnas in 1986/87].

Reaction to the following statements	Low	Medium	High
- 'present bad condition is the result of wrong deeds in the past life'	0.84	0.64	0.51
- 'what is written in the forehead (or fate) is never written off'	0.91	0.71	0.52
- 'one day God will give a happy day to all'	0.63	0.49	0.42
- 'present problems are created by God'	0.67	0.45	0.36
- 'religious work solves the present problems'	0.65	0.49	0.45
Total	0.74	0.55	0.43

(Source : Household Survey, 1987).

Table A.10: Reasons for not visiting the Government office (number and %age of respondents) [Begnas, 1987].

Reasons	Number of respondents
-'no suggestions are available'	79 (44.4)
-'location at far distance'	70 (39.9)
-'others'	29 (16.3)

(Source: Household Survey, 1987). Figures in brackets are percentages.

Table A.11: A generalized model of division of work by gender (as observed in Begnas in 1987).

General types of work	Male	Female	Children
Ploughing using oxen*	A	*	**
Digging fields	U	O	R
Terrace construction and maintenance	U	O	R
Levelling land	U	O	R
Hauling manure	O	U	R
Fetching fuelwood and fodder from forest	O	U	R
Fetching cut-grass	O	U	R
Feeding animals	O	U	R
Milking animals	O	U	**
Cleaning animal shed	O	U	R
Masonry and carpentry work	A	**	**
Planting seedling	R	U	**
Sowing seed	O	U	**
Uprooting seedlings	O	U	R
Irrigating fields	U	O	R
Weeding crops	O	U	R
Harvesting crops	U	U	R
Threshing	U	R	R
Hauling grains	O	U	R
Processing food grains	O	U	R
Fetching water	R	U	R
Storing food grains	R	U	R
Cooking, washing, house cleaning and maintenance	R	U	**
House repair	U	O	R
Child care	R	U	O
Tending animals	O	O	U

A = Always done (by) U = Usually done (by) O = Occasionally done (by) R = Rarely done (by)

* : Forbidden due to religious reasons **: Note performed

Note: This is just the illustration of the division of different types of works by age and gender. There are divisions within each type of activity mentioned here.

INDEX

A

B

C